Ana Carina Matos Silva

Ecotoxicological assessment of crude oil: two case studies

Ana Carina Matos Silva

Ecotoxicological assessment of crude oil: two case studies

Analysis of the toxicological response of microorganisms in a simulated spill environment

ScienciaScripts

Imprint

Any brand names and product names mentioned in this book are subject to trademark, brand or patent protection and are trademarks or registered trademarks of their respective holders. The use of brand names, product names, common names, trade names, product descriptions etc. even without a particular marking in this work is in no way to be construed to mean that such names may be regarded as unrestricted in respect of trademark and brand protection legislation and could thus be used by anyone.

Cover image: www.ingimage.com

This book is a translation from the original published under ISBN 978-613-9-61805-7.

Publisher:
Sciencia Scripts
is a trademark of
Dodo Books Indian Ocean Ltd. and OmniScriptum S.R.L publishing group

120 High Road, East Finchley, London, N2 9ED, United Kingdom
Str. Armeneasca 28/1, office 1, Chisinau MD-2012, Republic of Moldova, Europe
Printed at: see last page
ISBN: 978-620-7-70144-5

SUMMARY

I dedicate this work to my entire family, especially my mother, Màrcia, my father, Alberto, my grandparents and especially my aunt Gil (*in memoriam*) who will always be a great source of inspiration.

ACKNOWLEDGMENTS

I would like to thank my mother, father and sister, who with all their support and dedication made this journey possible, and my little dog, for welcoming me happily every day.

To my family, for their patience, understanding and for always encouraging me to grow.

To my ever-present friends (Nailinha, Celi, Italo, Samy, Gabi, Sil, Laise, Leticia, Monta) for providing me with wonderful moments of distraction.

Friends gained during my postgraduate studies (Bel, Andressa, Maria, Nuria, Daiane, Carine, Verònica, Alexandre) for all the incredible experiences, even in this short time together.

To Professor Jorge Alberto Triguis, for taking me on as an advisor, believing in the project and enriching me with huge doses of encouragement.

To Professor Olivia Oliveira, for all her help, and especially for adopting me as one of her children in this huge family and for always being willing to solve each of my problems in a simple and effective way.

To my friend and doctor, Icaro Thiago, for his encouragement, his presence, for believing in my potential, for his trust, for his help in the most critical moments, and for sponsoring me in this ongoing process of academic development.

To Professor Antônio Fernando Queiroz, for his commitment in making this project possible, together with the teachers and staff of POSPETRO.

To the entire NEA-LEPETRO family (especially Eduardo, Gisele, Jorginho, Marcos, Sara, Daniela, Claudia, Isabel, Karina, Cicero, Alex, Lula and Adriana) for supporting the old and new researchers in our group.

To the staff of the Rio Grande University Foundation (FURG) for providing lodging, transportation, food and security during the time I was carrying out my project in Rio Grande.

To CONECO, professors, researchers and technicians, for supporting the project, providing the most basic tests and guidance.

To Professor Gilberto Filmann, Professor Grasiela Pinho and Italo Castro for their help, good ideas and clarifications.

To Sanye for being by my side all the time (I'll never be able to thank her enough for her efforts) to Jûlia, Rodrigo and Natâlia, for the sightseeing and gastronomic tours, for the fun and the welcome.

To LAMEB and Professor Marlene for kindly lending me their monitoring equipment.

To the technicians of the Plasma Laboratory (IGEO/UFBA) and especially to doctoral student Antônio Bonfim for the liters of distilled water lent and for making his equipment available for part of the analyses.

To the Coordination for the Improvement of Higher Education Personnel - CAPES, for the grant.

To the project Geoenvironmental Diagnosis of Mangrove Zones and Development of Technological Processes Applicable to the Remediation of These Zones: Subsidies for an Impact Prevention Program in Areas with Potential for Petroleum Activities in the Southern Coastal Region of the State of Bahia (PETROTECMANGUE-BASUL), which provided resources that made this research possible.

To each of the individuals who had to be sacrificed in order to carry out this research.

Yes, I know. It may sound crazy, but I'm really going to unscrew them and throw them into the garden. Even if they can't fly, they'll stay among the flowers, the rightful place of paralyzed butterflies. I can no longer bear the thought of opening the window, lifting the glass and seeing them there: disguised as hinges

Rita Apoena

SUMMARY

In the event of an oil spill in mangrove areas, there is a great possibility that the oil will aggregate with the suspended particulate material, which could determine the bottom sediment as the final destination of the OSA (Oil-Suspended Particulate Aggregate). In many cases, this formation can represent a major risk to benthic and nectonic organisms. Ecotoxicity tests with these organisms have been widely used as one of the most accurate forms of biological monitoring of anthropogenic impacts on different ecosystems. This study evaluates the ecotoxicity of the OSA formation in mangrove simulations, using sediment samples collected along the Rio Pardo estuary, in the municipality of Canavierias, Bahia, through pilot-scale tests with simulation units, carried out with oil from the Campos basin. The procedure was carried out using acute and chronic exposure toxicology tests to determine the LC_{50} (50% lethal concentration) using the copepod *Nitokra sp.* and the microcrustacean *Artemia salina*. Based on the sensitivity tests, the toxicological effects exhibited in the bioassays were characterized and the levels of mortality (survival) were analysed. It was found that, for *Nitokra* sp., the scenario in which OSA was formed had less toxic potential (CL_{50} 70.71) than elutriate (CL_{50} 5.59), confirming its potential as an effective way of cleaning oil in water. For *Artemia salina,* the sediment concentration in an OSA formation simulation that showed the greatest toxic potential was 200 mg/L, with equal values for surface and bottom (CL_{50} 7.91%), while the concentration with the lowest toxic potential was 300 mg/L (CL_{50} 31.5) for surface samples. This study presents innovative procedures for assessing the ecotoxicity of OSA in simulated mangrove environments, demonstrating the potential suitability of *Nitokra sp.* and *Artemia salina* in laboratory tests.

Keywords: Ecotoxicology. *Nitokra sp. Artemia salina*. Mangrove sediment. Oil. Oil-Suspended Particulate Aggregate (OSA).

1   INTRODUCTION

The mangrove is a coastal interface ecosystem characteristic of estuarine zones in tropical and subtropical regions, subject to the oceanographic influence of tidal changes, dominated by typical plant species, associated with other components of flora and fauna, microscopic and macroscopic, adapted to a periodically flooded substrate, with biogeochemical adaptations to large variations in salinity. (SANTOS et al., 2012; QUEIROZ; CELINO, 2008). In this environment, although highly resilient and resistant, the granulometric constitution of the sediment allows pollutants to be retained and their toxicity is magnified, significantly affecting the integrity of the ecosystem. Petroleum hydrocarbons stand out as the most harmful compounds, capable of producing immediate damage to organisms and their damage will depend on both the characteristics of the oil (type, quantity, quality and state of weathering) and the seasonal variations in the prevailing climatic and tidal conditions (BRITO et al., 2009).

Considering the vulnerability of mangrove regions and various other coastal ecosystems, natural dispersion is highly desirable for oil slick removal, as it favors the dissolution of soluble or more volatile hydrocarbons, which end up being preferentially removed by evaporation, and increase the area of the oil/water interface available for biological activity, which can increase degradation, making it a faster process compared to alternatives for cleaning up areas subjected to oil spills (ITOPF, 2013).

In this respect, studies related to Oil Suspended Material Particulate Aggregates (OSAs), a complex that arises from the interaction between mineral particles and oil in an aqueous medium, have recently been very prominent and have been widely recognized as an effective cleaning method for treating oil in water (LEE et al., 2002; KHELIFA et al., 2007; WANG et al., 2013; SUN et al., 2009). Oil droplets are easily fragmented into sizes of a few micrometers, which leads to the rapid transfer of oil slicks from the sea surface to the water column by increasing their contact surface (LE FLOCH et al., 2002; STOFFYN-EGLI; LEE, 2002). Factors that influence the interaction of the incorporated oil with the solid phase include the character of the oil (viscosity, composition), the type of mineral (mineralogy, granulometry, organic matter content), the hydrodynamic energy of the environment and the salinity of the water (STOFFYN- EGLI; LEE, 2002; OWENS; LEE, 2003; SERGY et al., 2003).

The formation of these droplets can increase the concentration of hydrocarbons in solution, also increasing the bioavailability of these products, promoting not only an increase in natural degradation but also an increase in the toxicity of the oil due to exposure to compounds such as benzene, toluene and xylene. These compounds have considerable solubility in water compared to others, which makes marine organisms more vulnerable, since they absorb these contaminants through their tissues, gills, direct ingestion of water or contaminated food (ITOPF, 2013).

Marine organisms are often used in the toxicological assessment of sediments because of their association with both the liquid and solid phases of the environment, providing coherent responses to the different implications of each of these parts of the ecosystem (USEPA, 1991). On the other hand, zooplanktonic organisms are totally interdependent on the aquatic environment, making them suitable for toxicity tests in saline environments. Both habits are relevant to toxicity studies because they are a trophic link between marine habitat communities and higher chains (VEIGA; VITAL, 2002).

Toxicity tests can be defined as laboratory tests which, when carried out under specific and controlled experimental conditions, make it possible to estimate the toxicity of different substances, industrial effluents and environmental samples (water or sediment). Toxicological tests are based on exposing test organisms to different concentrations of the sample to be analyzed, and the toxic effects exhibited are observed, described and quantified. Describing the toxic effects of the substances analyzed on marine and estuarine organisms is essential for maintaining biodiversity (NASCIMENTO; SOUZA; NIPPER, 2002).

The first records of toxicity tests with aquatic organisms date back to 1920, with fish being used first. In later decades, the 1940s and 1950s, work in the field became more prominent and other methods were developed. Differences in experimental conditions were improved, and the first studies related to environmental toxicology included pesticide action tests on insects and water quality tests with cladocerans. In the 1940s and 1950s, studies began on the effects of chemical substances and residues on the environment, and with them the realization that these tests needed to be standardized (HUNN, 1989 apud NASCIMENTO; SOUZA; NIPPER, 2002).

In general, toxicity tests must be regulated by bodies that develop test protocols and technical standards. In Brazil, the body responsible for developing these protocols is the Brazilian Association of Technical Standards (ABNT). The São Paulo State Environmental Sanitation Technology Company (CETESB) also standardizes toxicity tests. Other international bodies such as the *Association Française de Normalisation* (AFNOR), the *American Society for Testing and Materials* (ASTM), the *American Water Work Association* (AWWA), the *Deutsches Institut fur Normung* (DIN), the *International Organization for Standardization* (ISO) and the *Organization for Economic Cooperation and Development* (OECD), have also focused on the development of toxicity testing protocols that define permissible toxicity limits with acceptable levels of uncertainty and that serve as a reference for regulatory bodies when making decisions (COSTA et al., 2008).

Toxicity tests are classified into acute and chronic. Chronic toxicity tests are used to estimate the toxic effects of substances under conditions of prolonged exposure to sub-lethal concentrations, i.e.

concentrations that do not limit the survival of organisms, but which in some way affect their biological functions in conditions such as mobility, reproduction, feeding or other metabolic functions. The results obtained in chronic toxicity tests are generally expressed as CENO (Concentration of Unobserved Effect) or CEO (Concentration of Observed Effect), but can also be expressed as CE50 (Average Effective Concentration in 50% of organisms) (ARAÙJO; ARAÙJO-CASTRO; SOUZA-SANTOS, 2006; KUSK; WOLLENBERGER, 2007; ARAUJO-CASTRO, 2008; ARAUJO-CASTRO et al, 2009; GARCIA, 2009; OLIVEIRA, 2011).

Acute toxicity tests are widely used to measure the possible effects of toxic substances on certain species over a short period of time compared to the life span of the test organism in question. It assesses the dose or concentration of a toxic agent that would produce a specific measurable reaction in a test organism over a short period of time, usually 24 to 96 hours. The main effect measured in acute toxicity studies is lethality or some other behavior that precedes it, such as immobility. Acute toxicity tests allow for EC50 (Effective Concentration) and LC50 (Lethal Concentration) values. The values of effective and lethal concentrations are always related to 50% of the organisms because these responses are more reproducible, can be estimated with a greater degree of reliability and are more significant for extrapolation to a population (COSTA et al., 2008).

Several studies have been carried out to determine the toxicity of petroleum hydrocarbons, however, some studies that relate the presence of oil in sediment and its toxicological response to different organisms are worth highlighting (CRUZ; TRIGÜIS; QUEIROZ, 2012; COSTA, 2010).

In oil spill simulation studies, toxicological analyses have shown that there are significant responses directly related to oil degradation (CRUZ; TRIGÜIS; QUEIROZ, 2012). Other tests have revealed that treatments such as sieving increase the toxicity of oil-contaminated sediment subjected to remediation treatments for copepods, significantly reducing fecundity for these organisms (COSTA, 2010). Appendix A provides a broader overview of recent studies on OSA and the organisms used in this study.

For Castro (2009), toxicological tests are one of the most precise forms of biological monitoring in mangrove environments subjected to anthropogenic impacts. For this reason, his work with three different hydrocarbon fractions is worth mentioning, where he stipulated the relationship between acute toxicity for microorganisms and the density of the oil, showing that the most pronounced toxicity varies from the lightest to the heaviest types of oil, due to the high volatility of the hydrocarbons, which proved to be more toxic because they had a higher concentration of aromatic compounds.

Based on sensitivity tests using the reference substance (Sodium Dodecyl Sulphate - DSS), possible toxicological and physiological effects were characterized in different bioassays and mortality levels (survival) were analyzed. This work presents innovative procedures for assessing ecotoxicity after the formation of OSA in mangrove environments, demonstrating the potential suitability of *Nitokra* sp. and *Artemia salina* in laboratory tests.

1.1 THEORETICAL BASIS

Brazil's extensive coastal areas have been constantly threatened by the impacts of the petrochemical sector, which concentrates more than half of its crude oil production and almost all of its natural gas production on offshore platforms (ZIOLLI, 2002).

Among coastal ecosystems, the mangrove stands out as one of the most productive in the biosphere. It is characteristic of estuarine zones in tropical and subtropical regions and is subject to the oceanographic influence of tidal changes. Since mangroves are located near the coast, where oil spills are most common, the ecosystem is constantly threatened by this type of pollutant (QUEIROZ; CELINO, 2008).

According to Gundlach and Hayes (1978), mangroves are classified as one of the most sensitive and vulnerable coastal ecosystems to oil spills. Oil can remain persistent in this ecosystem for years, which makes techniques to clean up or remove oil from these environments quite limited. For this reason, the application of dispersants to oil spills has been considered a technically viable option. Dispersants are defined as formulations designed to reduce the surface tension between oil and water, helping to disperse the oil in droplets in the aqueous medium (BRASIL, 2000).

Currently in Brazil, the use of dispersants is regulated by CONAMA Resolution No. 269 of September 14, 2000, which establishes, among other things, the criteria and restrictions for decision-making regarding the use of dispersants, although the use of chemical dispersants is not allowed in sheltered and sensitive coastal environments, such as mangroves. Considering the high sensitivity of coastal regions, the use of dispersants is carefully established and accepted only if it results in less environmental damage, when compared to the effect caused by a spill without any treatment, or used as an alternative or even additional option to mechanical containment and collection in the event that these response procedures are ineffective (BRASIL, 2000).

In addition to the high vulnerability of coastal regions, the use of dispersants is a controversial alternative, because despite the aim of minimizing the damage caused by the oil slick, many species are highly sensitive to its components, as well as to hydrocarbons that become bioavailable (GESAMP, 1993).

Unlike dispersion attenuated by surfactants, natural dispersion through the association of oil with suspended particulate material is highly desirable for removing oil stains, as it favors the dissolution of soluble or more volatile hydrocarbons, which end up being preferentially removed by evaporation, and increase the area of the oil/water interface available for biological activity, which can increase degradation, making it a faster process compared to alternatives for cleaning areas subjected to oil spills (PIMENTEL, 2007).

The copepod *Nitokra* sp. although not standardized in Brazil for toxicity tests, has been widely used in toxicological bioassays carried out in sediment because it has peculiar characteristics that favor its use as a test organism. These characteristics include its enormous potential for cultivation in laboratory conditions, its generalist nature in terms of sediment granulometry, its constant sensitivity to the reference substance (DSS), its close association with the sedimentary environment, its high abundance, its short life cycle and its high ecological importance. In addition, representatives with varied life stages can be obtained from cultures at any time of the year, are easy to grow in the laboratory and require minimal space and equipment to carry out the tests (FERRAZ et al., 2010; FURLEY et al., 2010; ZARONI et al., 2010; SOUSA et al., 2012).

The microcrustacean *Artemia salina* (LEACH, 1819) is used extensively as a test organism in ecotoxicological laboratory tests, mainly related to estuarine, marine and hypersaline environments. Among the intrinsic characteristics that make this organism widely used in laboratory tests are its wide range of tolerance to salinity (5-250) and temperature (6-25°C); short life cycle; high adaptability to environmental adversities and test and cultivation conditions; high fecundity (bisexual or parthenogenetic reproduction); small size; non-selective diet; high adaptability to a variety of nutrients, guaranteeing reliability, viability and cost-effectiveness of the tests (NUNES et al... 2006), 2006).

The scope of this work stems from the need to identify new techniques with methodological applicability for toxicological testing, with the aim of validating the formation of the Oil-Suspended Particulate Aggregate (OSA) as a highly effective mechanism for remediation in cases of hydrocarbon accidents in coastal environments.

1.2 BACKGROUND

According to Mançù and collaborators (2006), national and foreign companies have given priority to the management of environmental impacts in oil exploration and production in the state of Bahia, highlighting in their main initiatives the allocation of efforts and resources for the gradual construction of an adequate Environmental Management System, bearing in mind that the oil industry brings in the essence of its prospecting, extraction, production and service activities, the generation of waste that can pose risks to the environment.

Taking into account the gradual pace of oil exploration in coastal environments, a plausible management policy is needed for decision-making based on quantifying the risks of exposure to petroleum hydrocarbons and other compounds for contaminated areas.

Toxicology is a science whose object of study is the possible harmful effect caused by the interaction between a toxic agent and a biological system, with the main objective of preventing the appearance of this effect, i.e. providing criteria for safe conditions of exposure to these substances. Ecotoxicological studies are widely applicable to various situations of environmental impact caused by chemical compounds, representing a useful tool for environmental management of coastal areas affected by oil spill processes (PEDROZO et al., 2002).

In agreement with Castro (2009), it is considered that toxicological analyses make it possible to determine, based on simulations that integrate exposure and risk assessment methods, relevant subsidies to the decision-making process, mainly regarding the allocation of resources, the urgency of corrective actions, the need for remediation, acceptable remediation levels and applicable technological alternatives, justifying the scope of this project, which is mainly applicable to the management of contaminated coastal areas.

1.3 PREMISES AND HYPOTHESES

•   Oil-Suspended Particulate Aggregate (OSA) is an ecologically coherent natural formation that makes a unique contribution to oil dispersion;

•   ecotoxicological tests can provide answers about the degree of harmfulness of exposure to crude oil and the formation of the Oil-Suspended Particulate Aggregate (OSA);

•   *Artemia salina and Nitokra* sp. can be applied as test organisms in bioassays with petroleum components, responding coherently to the different concentrations to which they are exposed;

•   the formation of the Oil-Suspended Particulate Aggregate (OSA) is less harmful to organisms when compared to exposure to crude oil;

•   different sediment concentrations in mangrove simulation units, at the microscale level, exert different toxicity potentials for the microcrustacean *Artemia salina*.

2   OBJECTIVES

2.1 GENERAL OBJECTIVE

To verify the toxicological response of the harpacticoid copepod *Nitokra sp.* and the microcrustacean *Artemia salina* in acute toxicity assessment tests related to the formation of OSA in a simulated mangrove ecosystem at the microscale level, compared to exposure to crude oil.

2.2 SPECIFIC OBJECTIVES

•   Carry out acute toxicity tests from different simulated oil-contaminated environments, varying the sediment concentration, to determine the CL_{50} for *Nitokra* sp. and *Artemia salina* based on sensitivity tests;

•   to characterize the toxicological effect by defining the lethal levels of exposure after the formation of the Oil-Suspended Particulate Aggregate (OSA) on *Artemia salina* and *Nitokra* sp. in mangrove sediments;

•   to carry out a comparative analysis of the toxicity associated with exposure to the aqueous phase of a simulation in which the formation of Oil-Suspended Particulate Aggregate (OSA) occurred at different sediment concentrations, using direct exposure tests in microscale simulations;

•   correlate the physical and chemical parameters monitored with the possible effects of toxicological tests.

3   MATERIALS AND METHODS

The methodology used was based on toxicological tests with static tests without renewal, where it is possible to determine possible adverse effects of exposure to samples taken in tests simulating the formation of OSA and crude oil on the survival of the copepod *Nitokra* sp. and the microcrustacean *Artemia salina* (acute test). The acute tests with the liquid phase made it possible to estimate the percentage dilution value, which represents lethality or immobility as a statistically observable effect for the test organism.

The copepod *Nitokra* sp. was used as a test organism in acute tests for toxicological tests representative of sediment samples, due to its benthic habit being closely associated with sediment conditions. The microcrustacean *Artemia s.* was used for acute tests in microscale simulations of OSA formation in a mangrove environment.

Three toxicological exposure models were evaluated for each of the tests. The first model represents the exposure of organisms to the simulated environment with OSA formation, the second model is representative of exposure to the environment without OSA formation, and the third model corresponds to the control environment.

To carry out these tests, several stages were carried out, including the collection and storage of sediment, laboratory analysis and characterization of this sediment, cultivation and acclimatization of the organisms, pre-tests with reference substances and definitive tests.

Throughout the course of the project, constant bibliographic surveys were carried out in order to enrich the information on the tests, processes and techniques used and also with the aim of publishing the actions explained above.

3.1 SEDIMENT COLLECTION AND STORAGE

For each of the three proposed simulations, sediment was sampled at six points during low tide along the Rio Pardo, near the town of Canavieiras, Bahia (Figures 3.1 and 3.2), taking care to collect the sediment in the first five centimeters with a stainless steel spatula that had been previously decontaminated and treated with local water. The physical and chemical parameters of the sediment (pH, Eh, temperature, salinity and conductivity) were measured *in situ* using a multi-parameter probe.

Figure 3.1 - Satellite image of the study area indicating the six collection points along the Rio Pardo, municipality of Canavieiras, Bahia

Source: Google Maps/BatchGeo, 2013

The sediment was collected within the limits of a transect approximately four meters long where 20 portions of the mangrove substrate were randomly collected and homogenized in a glass tray (Figure 3.3), and duly conditioned.

The sediment was stored in glass containers (for characterization) and aluminium containers (for simulation) and kept in Styrofoam boxes with ice during transport, at a temperature of up to 4°C until it reached the Petroleum Studies Laboratory (LEPETRO/IGEO/UFBA). Further details and justifications for the specific choice of study area are given in Appendix B.

Figure 3.2 - Representative map of the study area showing the sampling points located along the Rio Pardo in the municipality of Canavieiras, Bahia.

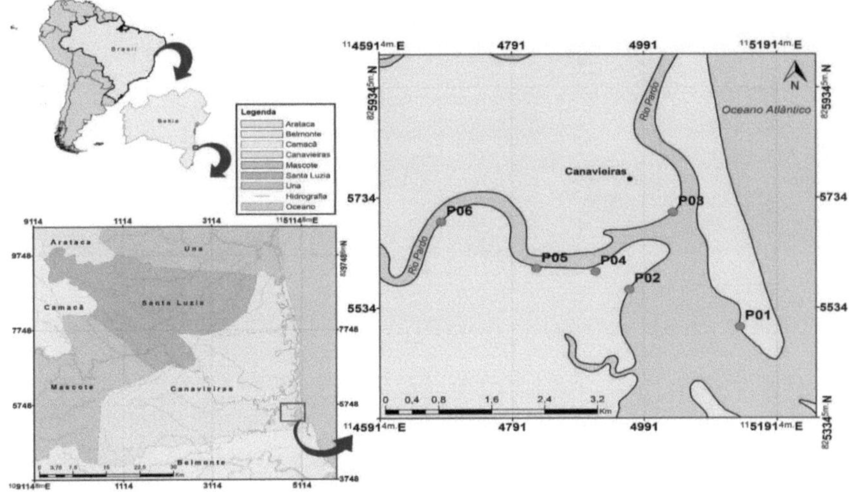

Source: The author, 2013

For characterization, the samples were freeze-dried using

They were then ground and disaggregated in a mortar and then homogenized and sieved through nylon fibre sieves in order to separate the fraction below 200µm, to remove the coarser fragments.

For the simulation, each sediment sample was calcined in a muffle furnace at 450°C for 6 hours to remove the organic matter, homogenized and sieved through a 100 µm mesh. This material was kept in a refrigerator at 4 °C in glass containers for a period of 24 hours until the sediment was distributed in the test containers for the bioassays to begin.

Figure 3.3 - Homogenization and conditioning of the sediment collected on the banks of the Rio Pardo in the municipality of Canavieiras, Bahia, for the OSA formation protocol. (A) Homogenizing the sediment in a glass container. (B) Storing the sediment in aluminium containers for simulation. (C) Storing the collected sediment in glass jars. (D) Thermal box for storing the collected sediment.

Source: The author, 2013

3.2 LABORATORY CHARACTERIZATION ANALYSES

The characterization of the oil and sediment matrices was carried out at the Petroleum Studies Laboratory (LEPETRO/IGEO/UFBA), and included analyses of sediment granulometry, APi grade (density) and viscosity/fluidity, HTP and HPA in the oil. Sediment analysis did not exceed 48 hours after collection. The techniques are explained in Table 3.1 and then described in more detail.

Table 3.1 - List of monitored parameters for the sediment and oil matrices used in the OSA simulation experiments, preceded by a description of the equipment/technique and its detection limit

Analysis	Equipment/Technique	Matrix	
		Sediment	Oil
pH	Portable pH meter (± 0.01 pH units)	x	
Eh	Portable Eh meter (± 0.01 Eh units)	x	
Temperature	Thermometer attached to the oximeter (± 0.5° C)	x	
Salinity	Portable conductivity meter (0.05%)	x	
Conductivity	Portable conductivity meter (0.05%)	x	
Dissolved oxygen	Multiparameter probe	x	
Granulometry	Laser diffraction (mod. Silas 1064)	x	
API grade/Density	DMA 5000 Density Meter		x
Viscosity / Fluidity	ASTM D97		x
Total Petroleum Hydrocarbons and Aromatic Polycyclics (HTP and HPA)	Liquid chromatography		x

In order to accurately characterize and describe the sediment samples to be used in the tests, laboratory analyses were necessary to provide a precise description of the most important characteristics of this sediment and the variables that interfere in the formation of OSA, making it possible to correlate them with the subsequent results.

It is important to note that before starting the analyses described here, the samples underwent a pre-treatment which included: freeze-drying to remove moisture; subsequent disaggregation, to disintegrate coarser particles; sieving and homogenization, which enabled the analyses to be carried out with the desired statistical fidelity.

a) Particle **size:** a Cilas 1064 laser diffraction particle size analyzer was used for the particle size analysis. The method consisted of pre-treating the sample with hydrogen peroxide (H_2O_2) to degrade the organic matter (EMBRAPA, 2009). After this stage, sodium hexametaphosphate ($Na_{(n+2)}P_n O_{(3n+1)}$)) was added and left for approximately 16 hours under agitation to avoid flocculation. The samples were classified by particle size (sand, silt and clay) according to the Folk and Ward classification (1957). Raw data is presented in Appendix C.

b) **Total Petroleum Hydrocarbons and Aromatic Polycyclics (HTP and HPA):** The organic compounds were determined using liquid chromatography extraction and subsequent analysis on a VARIAN CP 3800 gas chromatograph equipped with a DB-5 capillary column (30 m long, 0.25 mm internal diameter, 0.25 μm film thickness) and flame ionization detector (GC - FID). The chromatographic conditions were: injector temperature of 300 °C, initial oven temperature of 40 °C; 40 °C (for 2 minutes) in a ramp of 10 °C min^{-1} up to 300 °C for 12 minutes. Helium was used as the carrier gas at a flow rate of 1.0 mL min^{-1} and a separation ratio of 10:1 (BARRAGAN, 2012).

c) **Density / API grade:** This was carried out in accordance with ASTM 5002 and ISO 12.185 using a DMA 5000 Density Meter.

d) Pour Point: The sample was first placed in a solution of ethylene glycol and water and placed in an oven at a temperature of 60 °C to solubilize and be analyzed in the Cloud Pour Point equipment (HERZOG HCP 852), according to ASTM D97.

3.3 TOXICITY BIOASSAYS WITH *Nitokra* sp.

This section describes the use of *Nitokra* sp. in toxicity tests in simulations of environments that have been exposed to oil spills. Reference tests were carried out with a substance of known toxicity (Sodium Dodecyl Sulphate - DSS) to compose the control chart, in order to illustrate the regularization of the sensitivity of the cultivation of the organisms used for the tests, described in subtopic 3.3.2 Representative tests were also carried out on the aqueous fraction generated by simulating the formation of OSA at micro-scale level, as described in subtopic 3.3.3, and tests with elutriate from the homogenization of sediment and oil (subtopic 3.3.4).

It is important to note that each test was accompanied by analyses of the associated physico-chemical parameters (pH, dissolved oxygen, temperature and conductivity) at the beginning and end of the tests in order to control the basic conditions of exposure and support the interpretation of the results (NASCIMENTO; SOUZA; NIPPER, 2002).

3.3.1 Cultivation and acquisition of organisms

The acquisition of the organisms and the completion of this stage of the work was made possible through a partnership with the Laboratory of Organic Microcontaminants and Aquatic Ecotoxicology (CONECO) of the Federal University of Rio Grande (FURG), where the copepods were cultivated.

Cultivation is carried out in 1L erlenmeyer flasks containing around 750 mL of artificial seawater at salinities of 5, 15 and 30. The flask is covered with cotton wool and gas so as not to interfere with gas exchange and to prevent the water from evaporating (SOUSA et al., 2002).

The cultures were kept in an air-conditioned room at a temperature of $25 \pm 1°C$, in a greenhouse with a 12h light/dark photoperiod and without aeration. The water used for cultivation and partial water changes every 15 days was replenished with CoralLife artificial sea salt[®] . The algae of the species *Tetraselmis sp.*, *Isochrysis galbana* and *Thallasiosira weissflogii were* fed and supplemented with a diet prepared with TetraMin fish feed[®] . The diet was fed 2 to 3 times a week, adding 3 mL of fish feed to each flask.

This species was cultivated in the CONECO laboratory at three different salinities (5, 15 and 30), but in all the tests carried out in this work, the organisms at salinity 15 were used because they proved to be the best environmental conditions not only for the organism (FERRAZ et al., 2010) but also for the formation of the OSA (MOREIRA, 2014).

In order for these organisms to be used in the tests, they had to be selected from a small portion of the crop sieved through a 125 μm mesh so that only the adult individuals could be used, which were then counted and separated under a magnifying glass (NASCIMENTO; SOUZA; NIPPER, 2002).

3.3.2 Tests with reference substance Sodium Dodecyl Sulfate (DSS):

The sensitivity test was carried out by selecting adult males and non-ovulated females to be used as test organisms. The organisms were selected with the aid of a magnifying glass (optical stereomicroscope) using a petri dish filled with artificial brackish water, after they had been separated from the culture medium using a 125 μm mesh sieve.

The stock solution was prepared in 100 mL flasks where 10 mg of sodium dodecyl sulphate (NaC H$_{i225}$ SO$_4$) was added for every 100 mL of salt water at salinity 15, which is obtained by diluting

30g of artificial sea salt in 500 mL of water and volumizing to one liter.

Five concentrations were tested, with three replicates per concentration, with the following concentrations: 0 (control), 4, 6, 8, 10 and 12 mg.L^{-1} . Their exact dilution ratios are shown in Table 3.1. Ten adult males or non-ovulated females were exposed to each container containing 20 mL of the dilution solution.

The conditions for the test (Table 3.2) were maintained at 16 hours of light and 8 hours of dark, at a temperature of 25°C for 96 hours. The *endpoint* observed is CL50 of 96h. In other words, after a four-day exposure, the living and dead organisms in each test were counted using a microscope or magnifying glass. An acute test was also carried out with the water from the cultivation to maintain certification of the quality of the cultures (NASCIMENTO; SOUZA; NIPPER, 2002).

Table 3.1 - List of concentrations and dilution ratios of DSS in artificial brackish water for the preparation of test solutions for the acute toxicity test with *Nitokra* sp.

Concentrations (mg.L'1)	Stock solution (mL)	Brackish Water (mL)
12	12	88
10	10	90
8	8	92
6	6	94
4	4	96
0	-	100

Table 3.2 - Conditions for acute toxicity test (mortality) with *Nitokra* sp. for reference substance Sodium Dodecyl Sulphate

Parameters	Conditions
Temperature Salinity Photoperiod	25±1°C 15
Test system Test duration Volume of test solution Replicates per dilution Organisms per replicate	Natural (16h light/8h dark) Static 96h 10 mL 3 10 Adults
Age of organisms Effect observed	Lethality (CL50)
Validity of test	Minimum 70% survival in control

In this study, acute toxicological tests were carried out in simulation experiments, with elutriate derived from homogenized sediment and with the aqueous phase of the OSA formation simulation experiment at the micro-scale level. Two main scenarios were considered for comparison purposes, without the formation of Oil - Suspended Particulate Matter Aggregates and with the formation of Oil - Suspended Particulate Matter Aggregates. The acute tests were carried out using percentage dilutions of the elutriate (from the sediment) and the aqueous phase (from the OSA formation simulation experiment), with artificial brackish water at salinity 15.

3.3.3 Microscale simulation of the formation of Oil - Suspended Particulate Matter (OSA) Aggregates

Ninety erlenmeyer flasks were filled with 250 mL of artificial brackish water at salinity 15 and 0.175g of sediment was weighed for each, in order to respect the value of 0.7 g/L of sediment concentration, since according to Sun and Zeng (2009) this is the maximum concentration for the formation of OSA in particles of up to 16 µm.

The sediment was transferred to erlenmeyer flasks using a pisser filled with artificial brackish water. Each one was covered with aluminum foil, placed on a shaking table for one minute and left to stand overnight at 15°C in a darkened room. The following day, each flask spent another minute on the shaking table at 2.1 Hz (126 RPM) and then 0.05g (50 mg or 0.05 cm^3) of oil was added, taking the expected value of 50 mg. This was followed by a further three hours of stirring on a reciprocating shaking table at 126 RPM and another night's rest (MOREIRA, 2014).

The samples were vacuum-filtered to capture the aggregates formed on a cellulose acetate filter as described by Khelifa et al. (2007) and stored in darkened glass vials with the aid of a fine-tipped sieve containing brackish water at salinity 15.

It took 6 days of preparation to acquire the minimum total mass of contaminated sediment (12g) required for the chronic toxicity tests of the bottom sediment, where three replicates of each of the 6 percentage dilutions (100%, 50%, 25%, 12.5%, 6.25%, 1%) and a positive control (blank 0%) were evaluated.

3.3.4 Elutriate preparation protocol

Many sediment toxicity studies have evaluated the extraction of suspended sediment, called elutriates, as an aqueous representation of the sediment's characteristics. It is important to note that elutriates do not fully reflect the total toxicity of the sediment, since there are differences in the bioavailability of contaminants, so they are generally less toxic than the sediment itself.

The method for preparing the elutriate followed the USEPA (United States Environmental Protection Agency) standards described in the Technical Manual (USEPA, 2001) where the procedures described below are suggested.

One part sediment was combined with four parts brackish water at salinity 15, and the mixture was stirred vigorously for one hour on a shaking table. The compound was centrifuged for 10 minutes at 4,000 RPM in order to isolate the sediment from the aqueous phase and then used immediately after preparation. The preparation of the sediment for the tests and the elutriate followed the quantities set out in Table 3.2 below, which shows the ideal proportions obtained for the toxicity tests.

Table 3.2 - Proportion of sediment and oil for the preparation of the elutriate for the acute toxicological tests (ideal x real) using the copepod *Nitokra* sp.

Dilutions -	Clean Sediment (g)		Sediment + Oil (g)		Oil (g)
Percentages	Ideal	Real	Ideal	Real	
100%	20	20	-	-	20
50%	20	20	20	19,998	-
25%	30	29,720	10	9,938	-
12,5%	35	35,023	5	4,839	-
6,25%	37,5	37,507	2,5	2,730	-
1%	39,6	39,499	0,4	0,573	-

3.3.5 Acute toxicological tests with the aqueous phase of the microscale simulation of OSA formation

Acute toxicological tests were carried out on samples of the aqueous phase from the experiments simulating the formation of OSA in a mangrove environment at micro-scale level, using artificial brackish water at salinity 15 and the aqueous portion added during the preparation described in sub-item 3.3.3 (micro-scale simulation for the formation of the Oil-Suspended Particulate Aggregate), adding fractions collected from the surface using an automatic pipette in the following proportions (Table 3.3).

Table 3.3 - Values used for the aqueous phase of the micro-scale OSA formation protocol and artificial brackish water for acute toxicity tests using the copepod *Nitokra* sp.

Percentage Dilutions	Aqueous phase (mL)	Artificial brackish water (mL)
100%	15	-
50%	7,5	7,5
25%	3,75	11,25
12,5%	1,87	13,13
6,25%	0,94	14,6

1%   0,15   14,85

The acute toxicity test conditions for the simulation samples with the aqueous phase during the OSA simulation experiments are described in Table 3.3. For each test, only the OSA concentration of 700 mg/L of sediment was used for approximately 0.05g of oil, in seven dilutions, with three replicates for each dilution, with the following percentage proportions: 0% (control); 1%; 6.25%; 12.5%; 25%; 50%; and 100%.

Table 3.3 - Conditions for the acute toxicity test (mortality) with *Nitokra* sp. for the aqueous phase samples of the OSA formation simulation experiments

Parameters	Conditions
Temperature Salinity Photoperiod Test system Test duration Volume of test solution Replicates per dilution Number of dilutions Organisms per replicate Age of organisms Effect observed Validity of test	25±1°C 15 Natural (16h light/8h dark) Static 96h 15 mL 3 7 10 Adults Lethality (CL_{50}) Minimum 70% survival in control

3.3.6 Acute toxicology test with elutriate

The same methodology and experimental conditions were used for this test as for the aforementioned acute tests. Six percentage dilutions were tested, with three replicates per concentration, in the following proportions: 0% (control); 0.5%; 1%; 6.25%; 12.5%; 25%, in the exact dilution proportions shown in Table 3.4. Ten adult males or non-ovulated females were added to each container containing 15 mL of the dilution solution made from the elutriate.

The conditions for the test were kept similar to those shown in Table 3.3, respecting a 16-hour light and 8-hour dark period and a temperature of 25°C for 96 hours.

Table 3.4 - Values used for elutriate and artificial brackish water for acute toxicity tests with the copepod *Nitokra* sp.

Percentage Dilutions	Elutriate (mL)	Artificial brackish water (mL)
25%	3,75	11,25
12,5%	1,87	13,13
6,25%	0,94	14,6
1%	0,15	14,85
0,5%	0,075	14,925

3.4 TOXICITY BIOASSAYS WITH *Artemia salina*

This section discusses the methodology applied to the use of the micro-crustacean *Artemia salina* in acute toxicity tests to determine the lethal concentration of simulated samples at the micro-scale level of OSA formation in mangroves for 50% of the organisms exposed for a period of 48 hours (NASCIMENTO; SOUZA; NIPPER, 2002).

The procedures for the microscale tests were described in subtopic 3.3.3, in three different sediment concentrations: 50, 200 and 300 mgL^{-1} . For each simulation unit, 250 mL of artificial brackish water (salinity 15) and approximately 0.05g (50 mg or 0.05 cm^3) of oil were used (MOREIRA, 2014).

The collection of water for microscale testing is associated with the existence of a protocol designed to simulate an experimental spill and subsequent remediation treatment in a mangrove ecosystem. This protocol is part of a stage linked to the funding research project (Geoenvironmental Diagnosis of Mangrove Zones and Development of Technological Processes Applicable to the Remediation of These Zones: Subsidies for an Impact Prevention Program in Areas with Potential for Petroleum Activities in the Southern Coastal Region of the State of Bahia - PETROTECMANGUE-BASUL), which aims to develop mangrove simulation units at the microscale level to verify the formation of Oil-Suspended Particulate Aggregates (OSA) in mangrove areas, carrying out simulations of oil spills in order to assess the potential for OSA formation in mangrove environments.

For all the toxicological tests, the physico-chemical parameters (pH, redox potential, temperature and dissolved oxygen) were measured in each percentage dilution unit using portable probes.

3.4.1 Cultivation and acquisition of organisms:

The *Artemia salina* cysts were purchased from an aquarium and kept away from light in silica gel desiccators to ensure they were free from contaminants and to preserve their hatching rate.

They were then hatched in glass aquariums with a capacity of 1L with constant aeration, following the methodology suggested by Veiga and Vital (2002), not exceeding a concentration of 0.5 g/L of cysts. A salinity of 15 and an average temperature of 25°C were maintained in each tank.

48 hours after the cysts had hatched, the organisms were collected using a glass pasteur pipette to carry out the reference tests and toxicity tests, since at this stage they are suitable for testing because they are already beginning to filter, making them more sensitive and reducing the variability of the test (SORGELOOS et al., 1978).

3.4.2 Tests with reference substance Sodium Dodecyl Sulphate (DSS)

Three reference tests were carried out with the substance Sodium Dodecyl Sulphate ($NaC_{12}H_{25}SO_4$), to assess the sensitivity range of the organisms used in the test and to draw up the toxicity control chart. The samples were diluted in volumetric flasks to concentrations of 0, 9, 12, 16, 21, 27 and 35 mg/L.

Each dilution unit was carried out in triplicate with glass test vials containing 10 mL of the solution and ten organisms. The containers were placed on a tray and covered to minimize evaporation losses.

Table 3.4 illustrates the test conditions, which were maintained at a photoperiod of 12 hours light and 12 hours dark and a temperature of 25°C for 48 hours. The *endpoint* observed was the CL50 of 48 hours, i.e. live and dead organisms were counted after two days in each test using a counting chamber.

Table 3.4 - Conditions for acute toxicity test (mortality) with *Artemia salina* for reference substance Sodium Dodecyl Sulphate

Parameters	Conditions
Temperature Salinity Photoperiod Test system Test duration Volume of test solution Replicates per dilution Organisms per replicate Age of organisms Effect observed Validity of test	25±1°C 15 Natural (12h light/ 12h dark) Static 48h 10 mL 3 10 Levels II and III Lethality (CL50) Minimum 70% survival in control

3.4.3 Acute toxicological tests with the aqueous phase of the microscale simulation of OSA formation

The procedure for the acute toxicological tests with *Artemia salina* relating to the microscale simulation of the formation of the Oil-Suspended Particulate Aggregate (OSA) followed a similar route to that described in section 3.3.3. The erlenmeyer flasks were filled with 250 mL of artificial brackish water at salinity 15 and sediment was added in order to respect the concentrations of 0.05,

0.3 and 0.5 g/L. In addition to the flask containing different concentrations of sediment, the same process was carried out for a positive control (blank) containing only brackish water.

The sediment was transferred to erlenmeyer flasks with the help of a pisser filled with artificial brackish water, and each one was covered with aluminum foil, placed on a shaking table for one minute and left to rest overnight at a temperature of 15°C in a darkened room.

The following day, each of the flask was shaken for another minute at 2.1 Hz (126 RPM) and then 0.05g (50 mg or 0.05 cm^3) of oil was added. This was followed by a further three hours of stirring on a reciprocating shaker table at 126 RPM and another overnight rest.

For each sediment concentration tested, three replicates of each of the 5 percentage dilutions (100%, 50%, 25%, 12.5%, 1%) and a positive control (blank 0%) were evaluated. Surface and bottom samples were acquired for this test. The conditions for the microscale acute toxicity tests were similar to those described in Table 4.

3.5 ANALYSIS OF RESULTS

All the data obtained was subjected to statistical treatment using Excel 2010 and the Statistica 8 package. These programs were used to perform descriptive statistics, check the normality of the data and make graphs representing the physical-chemical characterization of the mangrove sediment, the results of the toxicological tests on the experimental units and the monitoring of the physical-chemical parameters of these units

The data produced by the acute tests with microscale and mesoscale simulations and with reference substances were expressed as adult mortality and survival data (CL_{50}) using the mathematical Probit method or Trimmed Spearman-Karber Method as recommended by the USEPA (2002) and associated with 95% confidence intervals. The tests with sediment and contaminant produced responses related to mortality (or survival) and fecundity (number of nucellus, copepodites and eggs) which will be given for each test in comparison with the control group.

The results were analyzed as a direct function of the type of test (acute) and it was possible to estimate lethal concentrations for 50% of the organisms. The use of these analyses makes it possible to estimate the concentration that causes a 50% reduction in the reproduction of the animals exposed to the different concentrations, in relation to the production observed in the control group.

Data analysis was interpreted by means of variance and comparison of the results of the different dilutions in relation to the control group. Pearson's linear correlation analyses were carried out between the toxicological and physico-chemical variables measured during the experiment. These correlations were carried out after checking the normality of the variables and the homogeneity of the variances. If these assumptions were not met, Spearman's correlation was used. Principal component analysis was also used to explain the variance between the groups.

The results of these tests allowed samples to be classified as toxic if the survival (or mortality) and/or fecundity of the sample was significantly different from the parameter obtained with the control and, necessarily, if there was a minimum difference of 20% between the mean obtained for the control and the substance tested (according to the principle of minimum significant difference). If the result did not meet these prerequisites, the sample was considered non-toxic (NASCIMENTO; SOUZA; NIPPER, 2002).

4 COMPARISON OF THE SENSITIVITY OF THE COPEPOD *NITOKRA* SP. TO TOXICOLOGICAL TESTS IN ESTUARINE SEDIMENT DURING THE FORMATION OF OIL-PARTICULATE-SUSPENDED-MATERIAL (OSA) AND ELUTRIATE AGGREGATES FROM SEDIMENT AND HOMOGENIZED CRUDE OIL

Summary:

In the event of an oil spill in mangrove areas, there is a great possibility that the oil will aggregate with the suspended particulate material, which could determine the bottom sediment as the final destination of the OSA (Oil-Suspended Particulate Aggregate). In many cases, this formation can represent a major risk to benthic organisms. Ecotoxicity tests with these organisms have been widely used as one of the most accurate forms of biological monitoring of anthropogenic impacts on this ecosystem. This study evaluates the ecotoxicity during the formation of OSA in mangrove sediments collected along the Pardo River estuary, in the municipality of Canavieiras, Bahia, by means of pilot-scale (microscale) tests. The procedure was carried out using acute exposure toxicology tests to determine the CL_{50} (50% lethal concentration) using the benthic copepod *Nitokra sp.* For comparison purposes, two scenarios were carried out, the first considering the aqueous phase of the protocol for simulating the formation of OSA on a microscale, and the second, the elutriate from sediment and oil homogenized in different fractions. It was found that the scenario in which OSA was formed had much less toxic potential (CL_{50} 70.71) than the elutriate formed from the percentages of sediment homogenized with crude oil (CL_{50} 5.59), confirming its potential as an effective form of cleaning for the treatment of oil in water.

4.1 INTRODUCTION

Mangroves are an interface ecosystem characteristic of estuarine zones in tropical and subtropical regions, subject to the oceanographic influence of tidal changes (SANTOS et al., 2012; QUEIROZ; CELINO, 2008). In these environments, the granulometric constitution of the sediment allows pollutants to be retained, increasing their toxicity and significantly affecting the integrity of the ecosystem. Petroleum hydrocarbons stand out as harmful compounds, capable of producing immediate damage to organisms and this damage will depend on both the characteristics of the oil (type, quantity, quality and state of weathering) and the seasonal variations in the prevailing climatic and tidal conditions (BRITO et al., 2009).

Considering the vulnerability of mangroves and other coastal ecosystems, natural dispersion is highly desirable for removing oil slicks, as it favors the dissolution of soluble or more volatile hydrocarbons, which are preferentially removed by evaporation, and increases the area of the oil/water interface available for biological activity, which can also increase degradation, making it a faster process compared to alternatives for cleaning up areas subjected to oil spills (ITOPF, 2013).

In this respect, there has recently been a great deal of emphasis on the Oil-Suspended Particulate Aggregate (OSA), a complex that arises from the interaction between mineral particles and oil in an aqueous medium and stands out because it has been widely accepted as an effective form of cleaning for the treatment of oil in water (LEE et al., 2002; KHELIFA et al., 2007; WANG et al., 2013; SUN et al., 2009). Oil droplets are easily fragmented into sizes of a few micrometers, which leads to the rapid transfer of oil slicks from the sea surface to the water column, increasing their contact surface (LE FLOCH et al., 2002; STOFFYN-EGLI; LEE, 2002).

The formation of these droplets can increase the concentration of hydrocarbons in solution, also increasing the bioavailability of these products, promoting not only an increase in natural degradation but also an increase in the toxicity of the oil due to exposure to compounds such as benzene, toluene and xylene. These compounds have considerable solubility in water compared to others, which makes marine organisms more vulnerable since they absorb these contaminants through their tissues, gills, direct ingestion of water or contaminated food (ITOPF, 2013).

The copepod *Nitokra* sp. although not standardized in Brazil for toxicity tests, has been widely used in toxicological bioassays carried out in sediment because it has peculiar characteristics that favor its use as a test organism. These characteristics include its enormous potential for cultivation in laboratory conditions, its generalist nature in terms of sediment granulometry, its constant sensitivity to the reference substance (DSS), its close association with the sedimentary environment, its high abundance, its short life cycle and its high ecological importance. In addition, representatives with varied life stages can be obtained from cultures at any time of the year, are easy to grow in the laboratory and require minimal space and equipment to carry out the tests (FERRAZ et al., 2010; FURLEY et al., 2010; ZARONI et al., 2010; SOUSA et al., 2012).

The aim of this study is to evaluate the sensitivity of the copepod *Nitokra* sp. in acute toxicological tests with static tests without renewal where it is possible to determine possible adverse effects of exposure to samples taken in tests where the formation of OSA was simulated in a microscale experimental protocol, as well as to the elutriate formed from the homogenization of percentage fractions of crude oil and sediment.

4.2 MATERIALS AND METHODS

In this study, acute toxicological tests were carried out in simulation experiments with sediment and its aqueous phase, considering two main scenarios: without the formation of Oil - Suspended Particulate Matter Aggregates and with the formation of Oil - Suspended Particulate Matter Aggregates. The acute tests were carried out in two main scenarios: the elutriate (from the sediment mixed with the oil) and the aqueous phase (from the OSA formation simulation experiment).

In parallel with each test, tests were carried out with a reference substance (Sodium Dodecyl Sulphate - DSS) to draw up a control chart with the aim of illustrating the regularization of the sensitivity of the cultivation of organisms used for the tests. Each test was accompanied by analyses of the associated physico-chemical parameters (pH, dissolved oxygen, temperature and conductivity) at the beginning and end of the tests in order to control the basic conditions of exposure and subsidize the interpretation of the results (NASCIMENTO; SOUZA; NIPPER, 2002).

The data produced by the acute tests with microscale and mesoscale simulations and with reference substances were expressed as mortality and survival data for adult individuals (CL_{50}) using the mathematical Probit method or Trimmed Spearman-Karber Method, as recommended by the USEPA (2002), associated with 95% confidence intervals. The sediment and contaminant tests produced responses related to mortality (or survival) which will be given for each test in comparison with the control group.

4.2.1 Sediment collection and storage

Sediments were sampled at six points during low tide along the Rio Pardo, near the municipality of Canavieiras (South Coast of the State of Bahia - Brazil), taking care to collect the sediments in the first five centimeters with a stainless steel spatula that had been previously decontaminated and treated with local water. The sediment was collected within the limits of a transect approximately four meters long where 20 portions of the mangrove substrate were randomly collected. The material was homogenized in a glass tray and stored at a temperature of up to 4 °C. The sediment's physical and chemical parameters (pH, redox potential, temperature, salinity, conductivity, dissolved oxygen, total solids and turbidity) were measured *in situ* using a multi-parameter probe.

To carry out the microscale simulations of OSA formation, each sediment sample was calcined in a muffle furnace at 450 °C for 6 hours, homogenized, sieved through a 100 μm mesh, kept in a refrigerator at 4 °C in glass containers for a period of 24 hours, until the sediment was distributed in the test containers for the bioassays to begin.

4.2.2 Oil and sediment characterization

The characterization of the oil and sediment matrices was carried out at the Petroleum Studies Laboratory (LEPETRO/IGEO/UFBA), and included analyses of sediment granulometry, API grade (density) and viscosity/fluidity, HTP and HPA in the oil. The sediment analyses were carried out within 48 hours of collection.

In the laboratory, after freeze-drying, sieving (2 mm) and homogenizing the samples, particle size analyses were carried out using a Cilas 1064 Laser Diffraction Particle Analyzer with sample pre-treatment according to Embrapa (2009).

Crude oil from the Campos basin was used for the simulations, and its density was measured six times with a densitometer (DMA 5000 Density Meter) at 15°C. Its viscosity was also determined three times under the same temperature conditions using a rotational viscometer (Haake Viscotester VT500).

The distribution of hydrocarbons was also used to characterize the oil by gas chromatography coupled to a Varian CP 3800 flame ionization detector (GC-FID), equipped with a 60 m long, 0.25 mm internal diameter DB5 capillary column, a 0.25 μm thick stationary phase and a flow of carrier gas (helium).

4.2.3 Cultivation and acquisition of organisms

The culture of the copepod *Nitokra* sp. was carried out in 1L erlenmeyer flasks containing about 750 mL of artificial seawater at salinity 15. The erlenmeyer flasks were covered with cotton and gauze so as not to interfere with gas exchange and prevent water evaporation and kept in an air-conditioned room at a temperature of 25 ± 1°C, in a greenhouse with a 12h light/dark photoperiod and without aeration (NASCIMENTO; SOUZA; NIPPER, 2002).

In order for these organisms to be used in the tests, they had to be selected from a small portion of the culture which was sieved through a 125 μm mesh so that only the adult individuals could be used, which were then counted and separated under an optical stereomicroscope.

4.2.4 Tests with reference substance Sodium Dodecyl Sulphate (DSS)

The stock solution was prepared in 100 mL flasks where 10 mg of sodium dodecyl sulphate (NaC H_{1225} SO_4) was added for every 100 mL of salt water at salinity 15, which is obtained by diluting 30g of artificial sea salt in 500 mL of water, volumized to one liter.

Five concentrations were tested, with three replicates per concentration, with the following concentrations: 0 (control), 4, 6, 8, 10 and 12 $mg.L^{-1}$. Ten adult males or non-ovulated females were exposed to each container containing 20 mL of the dilution solution.

The conditions for the test were maintained respecting the photoperiod of 16h of light and 8h of dark, temperature of 25°C for 96h. The *endpoint* observed is the CL_{50} of 96h. In other words, after an exposure of four days, the living and dead organisms in each test were counted using a microscope or magnifying glass. An acute test was also carried out with the water from the cultivation to maintain certification of the quality of the cultures (NASCIMENTO; SOUZA; NIPPER, 2002).

4.2.5 Microscale simulation of oil-material aggregate formation

Suspended Particulates (OSA)

Ninety erlenmeyers were filled with 250 mL of artificial brackish water at salinity 15 and 0.175g of sediment was weighed for each one, in order to respect the value of 0.7 g/L of sediment concentration, since according to Sun & Zeng (2009) this is the maximum concentration for the formation of OSA in particles of up to 16 μm.

The sediment was transferred to erlenmeyer flasks using a pisser filled with artificial brackish water. Each one was covered with aluminum foil, placed on a shaking table for one minute and left to rest overnight at 15°C in a darkened room. The following day, each flask spent another minute on the shaking table at 2.1 Hz (126 RPM) and then 0.05g (50 mg or 0.05 cm^3) of oil was added, taking the expected value of 50 mg. This was followed by another three hours of stirring on a reciprocating shaking table at 126 RPM and another night's rest (MOREIRA, 2014).

The samples were vacuum-filtered to capture the aggregates formed on a cellulose acetate filter as described by Khelifa et al. (2007) and transferred from the filters to darkened glass vials with the help of a fine-tipped sieve containing brackish water at salinity 15.

It took six days of preparation to acquire the minimum total mass of contaminated sediment (12g) required for the toxicity tests, where three replicates of each of the six percentage dilutions (100%, 50%, 25%, 12.5%, 6.25%, 1%) and a positive control (blank 0%) were evaluated.

4.2.6 Elutriate preparation

The method for preparing the elutriate followed the USEPA (United States Environmental Protection Agency) standards described in the Technical Manual (USEPA, 2001) by combining one part sediment with four parts brackish water at salinity 15. The mixture was vigorously stirred for one hour on a shaking table. The compound was centrifuged for 10 minutes at 4,000 RPM in order to isolate the sediment from the aqueous phase and then used immediately after preparation. The preparation of the sediment for the tests and for the elutriate followed the quantities set out in Table 4.1 below, which shows the ideal proportions and those obtained for the toxicity tests.

Table 4.1 - Proportions of sediment and oil used for acute toxicology preparation   of the elutriate to   the tests (ideal x real) using the copepod *Nitokra* sp.

Dilutions	Clean Sediment (g)		Sediment + Oil (g)		Oil(g)
Percentages	Ideal	Real	Ideal	Real	

100%	20	20	-	-	20
50%	20	20	20	19,998	-
25%	30	29,720	10	9,938	-
12,5%	35	35,023	5	4,839	-
6,25%	37,5	37,507	2,5	2,730	-
1%	39,6	39,499	0,4	0,573	-

4.2.7 Acute toxicological tests:

Acute toxicological tests were carried out on samples of the aqueous phase from the experiments simulating the formation of OSA in a mangrove environment at micro-scale level, using artificial brackish water at salinity 15 and the aqueous portion added during the experiment simulating the formation of OSA at micro-scale, adding fractions collected from the surface using an automatic pipette in the following proportions (Tables 4.2 and 4.3).

For each test, an OSA concentration of 700 mg/L of sediment was used for approximately 0.05g of oil, in seven dilutions, with three replicates for each dilution, with the following percentage proportions: 0% (control); 1%; 6.25%; 12.5%; 25%; 50%; and 100%. For elutriate, eight percentage dilutions were tested, with three replicates per concentration, in the following proportions: 0% (control); 0.5%; 1%; 6.25%; 12.5%; 25%, 50%, 100%. Ten adult males or non-ovulated females were added to each container containing 15 mL of the dilution solution made from the elutriate.

The salinity of the water was monitored using a portable refractometer and the physico-chemical parameters of all the tests were measured using a portable pH meter with an accuracy of ±0.01 pH units; O.D. with a portable digital microprocessor meter with an accuracy of ±0.05%; temperature with a thermometer coupled to an oximeter with an accuracy of ±0.05 °C and conductivity with a portable digital conductivity meter with an accuracy of ±0.05%. The conditions for the acute toxicity test are described in Table 4.1.

Table 4.2 - Values used for the aqueous phase of the OSA formation simulation test and artificial brackish water for acute toxicity tests using the copepod *Nitokra* sp.

Percentage Dilutions	Aqueous Phase (mL)	Artificial brackish water (mL)
100%	15	-
50%	7, 57,5	
25%	3,7511	,25

12,5%	1,8713	,13
6,25%	0,9414	,6
1%	0,1514	,85

Table 4.3 - Values used for elutriate and artificial brackish water for acute toxicity tests using the copepod *Nitokra* sp.

Percentage Dilutions	Elutriate (mL)	Artificial brackish water (mL)
100%	15	-
50%	7,5	7,5
25%	3,75	11,25
12,5%	1,87	13,13
6,25%	0,94	14,6
1%	0,15	14,85
0,5%	0,075	14,925

Table 4.1 - Conditions for the acute toxicity test (mortality) with *Nitokra* sp. for the aqueous phase samples of the OSA formation simulation experiments

Parameters	Conditions
Temperature Salinity Photoperiod Test system Test duration Volume of test solution Replicates per dilution Organisms per replicate Age of organisms Effect observed Validity of test	25±1°C 15 Natural (16h light/8h dark) Static 96h 15 mL 3 10 Adults Lethality (CL50) Minimum 70% survival in control

4.3 RESULTS AND DISCUSSION

4.3.1 Oil and sediment characterization

The results relating to the characterization of the sediment and oil used in this study will be described here, taking into account the field and sediment collection stage, which includes the measurement of physico-chemical parameters, as well as the laboratory analyses carried out. These results demonstrate the need to characterize the sediment used in the tests in order to control the experimental conditions evaluated. Table 4.4 shows the geographic coordinates and physico-chemical parameters of the collection points in the mangrove substrate of the study region.

The coordinates are directly related to the sampling points, as can be seen in Table 4.4, associated with the collection times described therein, which obeyed the restriction of collection at low tide (around 0.3m). The values for pH, Eh, temperature, salinity, conductivity, dissolved oxygen, total solids and turbidity are shown graphically below.

Table 4.4 - Geographical coordinates and descriptive physico-chemical parameters at the collection points on the banks of the Rio Pardo in the municipality of Canavieiras, south coast region of the state of Bahia

Points	Coordinates	Time	pH	Eh (mV)	Temp. (°C)	Salt.	Condt. (S/m)	O.D. (mL/L)	S.T.D. (mg/L)	Turbidity (NTU)
	Physico-chemical parameters (*in situ*)									
P1	S 15° 41' 58" W 38° 55' 54"	07:35	9,91	45	26,56	30	30,8	15,2	17,4	121,0
P2	S 15° 41' 36" W 38° 56' 21"	08:57	8,55	163	29,62	20	35,7	18,3	21,8	26,6
P3	S 15° 40' 52" W 38° 56' 24"	09:22	5,93	325	30,02	12	23,4	17,66	14,5	30,6
P4	S 15° 41' 26" W 38° 57' 05"	10:03	7,89	245	30,82	6	15,5	21,04	9,61	49,1
P5	S 15° 41' 25" W38° 57' 35"	10:56	7,61	240	30,37	6	14,8	19,63	9,16	99,2
P6	S 15° 40' 58" W 38° 58' 25"	11:47	7,91	222	30,08	5	12,3	18,95	7,62	179

Legend: pH = Hydrogen Potential; Eh = Reduction Potential; Temp. = Temperature; sal. = salinity; Condt. = Conductivity; O.D. = Dissolved Oxygen; S.T.D. = Total Dissolved Solids; NTu = Nephelometric Turbidity Units

Figure 4.1 - Graphical representation of pH values along the Rio Pardo in the municipality of Canavieiras, southern coastal region of the state of Bahia at the different sampling points

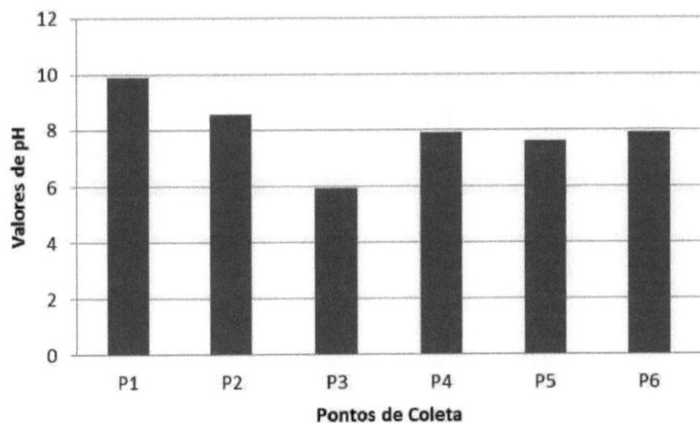

The pH values were neutral to slightly basic (with the exception of point 3, which was relatively acidic), with little variation between the six sampling points, with a maximum recorded at P1 (9.91) and a minimum recorded at P3 (5.93), which may be an indication of the uniformity of characteristics between these sites, a consequence of the similar contributions of marine and fresh waters at the sampling points (Figure 4.1).

With regard to Eh values, there was a wide variation between the maximums and minimums observed at points 3 and 1 (325.0 mV) and 5 (45.0 mV), respectively, all of which show a significant oxidizing characteristic, which represents the instability of this environment during the measurement of this parameter (Figure 4.2).

Figure 4.2 - Graphical representation of Eh values (in mV) along the Rio Pardo in the municipality of Canavieiras, south coast region of the state of Bahia at the different sampling points

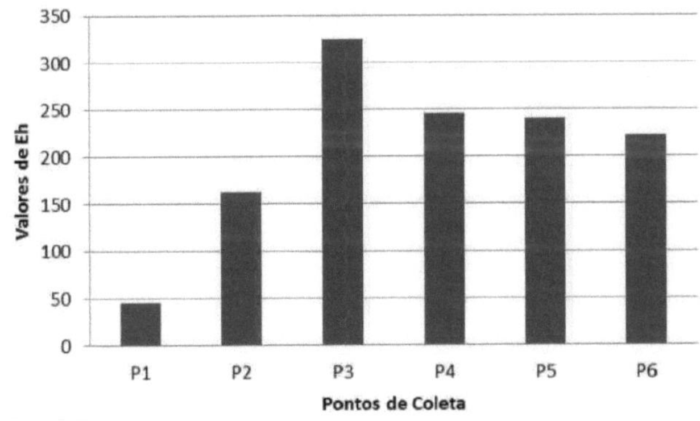

The variation in temperature did not exceed a difference of 4.2 °C between the points, with the maximum value recorded at P4 (30.8 °C) and the minimum value seen at P1 (26.5 °C). This variation in temperature may be related to the influence of climatic conditions depending on the times and places of collection, integrated with the intrinsic characteristics of the points sampled

(Figure 4.3).

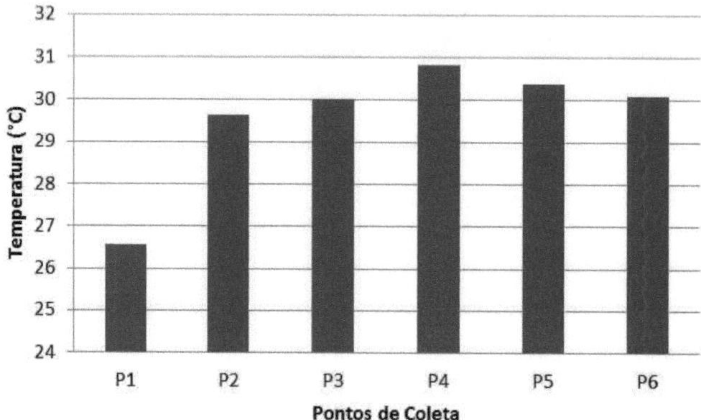

With regard to salinity, as expected for an estuary environment, it remained relatively high at the first three points (30, 20, 12) and decreased significantly at the last three (6, 6 and 5, respectively). In this case, it is important to note that there is a gradual reduction in values, where the highest indices were found at sampling points that are influenced by salinity downstream and the lowest upstream of the river (Figure 4.4). This behavior can be explained by the penetration of the tide, which causes samples collected in mangroves that are closer to the mouth of the estuary to have higher salinities.

The parameters pH, Eh and salinity are temperature-dependent, increasing and decreasing in a non-linear way according to temperature variations. All of these factors, together or in isolation, contribute to the cations that are sorbed to the particles that make up the mangrove substrates becoming bioavailable (QUEIROZ; CELINO, 2008). Therefore, the results of the physico-chemical parameters at the different sampling stations are crucial for correlating their possible effects as influences on the toxicity relationship found in the sediments.

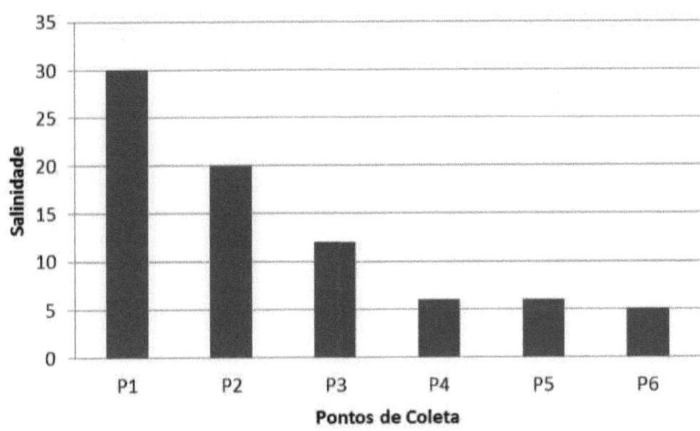

Dissolved oxygen values ranged from a maximum of 21.04 mL/L to a minimum of 15.20 mL/L (Figure 4.5). This variable is influenced by the composition of the sediments, the temperature, the salinity of the water and hydrodynamic factors. It offers indications of the natural conditions of the environment and makes it possible to detect possible environmental impacts such as eutrophication and organic pollution. Carbon dioxide, molecular oxygen, nitrite and nitrate ions and the water itself are the main sources of dissolved oxygen (BAUMGARTEN et.al., 1996).

It can be seen that dissolved oxygen had a behavioral pattern of values very similar to temperature. The temperature values at the different points can be considered a determining factor for the distribution of dissolved oxygen values, since temperature determines the solubility of gases (FIORUCCI, 2005).

With regard to the values found for conductivity (Figure 4.6), a wide variation was observed, with the highlight being the maximum value found in P2 (35.7 mS). It is possible to observe an inverse relationship between the behavior of Eh compared to conductivity, which can be interpreted from the perspective of the same criterion since the presence of dissolved ions can have a huge influence on the electrical conductivity values recorded (QUEIROZ; CELINO, 2008).

Figure 4.5 - Graphical representation of dissolved oxygen values (in mL/L) along the Rio Pardo in the municipality of Canavieiras, south coast region of the state of Bahia, at the different sampling points

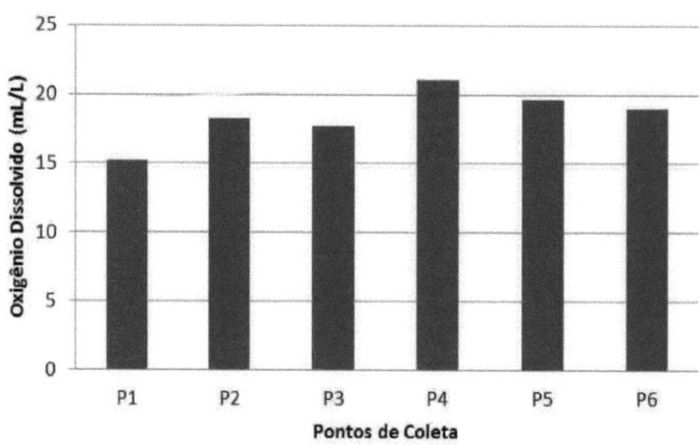

Figure 4.6 - Graphical representation of conductivity values (in S/m) along the Rio Pardo in the municipality of Canavieiras, south coast region of the state of Bahia, at the different sampling points

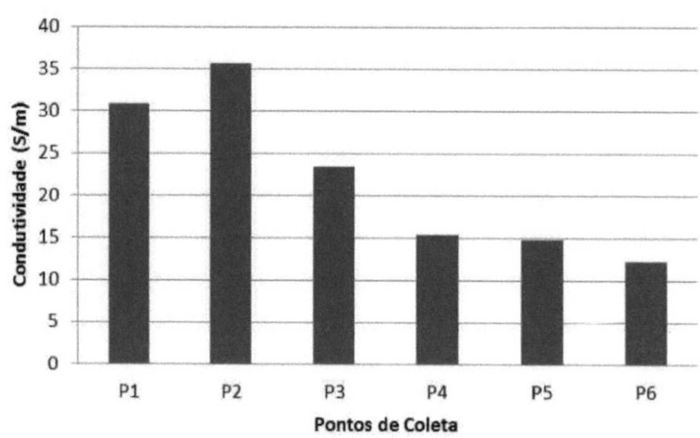

The highest value for total dissolved solids was found in P2 (21.8 mg/L) and the lowest in P6 (7.6 mg/L) (Figure 4.7), so the total solids values were directly proportional to those found for conductivity. Sampaio and colleagues (2007) had already carried out studies showing that there is a strong linear correlation between these two parameters, which can be explained by the fact that they are in solution and in a colloidal state (BRAILE; CAVALCANTI, 1993).

Figure 4.7 - Graphical representation of total dissolved solids values (in mg/L) along the Rio Pardo in the municipality of Canavieiras, south coast region of the state of Bahia, at the different sampling points

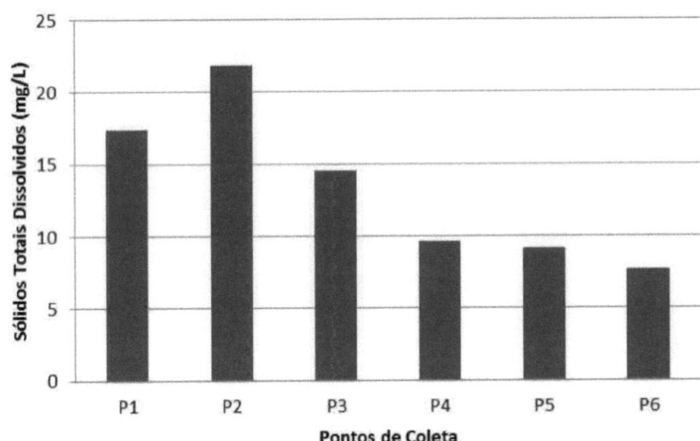

For turbidity, the highest value was found in P6 (179.0 NTU) followed by P1 (121.0 NTU) and the lowest value was recorded in P2 (20.6 NTU) (Figure 4.8). According to Pina et al. (2003), the high turbidity of the estuary is a consequence of the high tidal flow, the river load, the large expanse of intertidal areas and the high susceptibility to wave generation which intensifies resuspension.

Each of these values shows a set of environmental conditions in the study area that would enable and sometimes mitigate the formation of OSA in these locations, considering the factors that influence the formation of these aggregates in a natural environment.

Figure 4.8 - Graphical representation of turbidity values (in NTU) along the Rio Pardo in the municipality of Canavieiras, south coast region of the state of Bahia, at the different sampling points

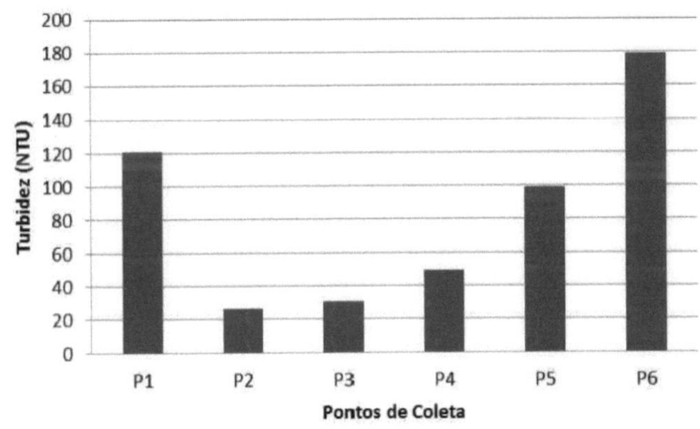

With regard to the granulometric profile of the collection points, the averages for the values found in the analyses show a predominance of fractions considered to be very fine sand (62.5 - 125 µm) and silt (3.9 - 62.5 µm) (Figure 4.9) according to the granulometric classification proposed by the Krumbein scale (1951). Considering that, according to Sun & Zeng (2009), granulometry is a determining factor in the formation and fate of the OSAs formed and that OSA formation is attenuated in sediment with a granulometry size of less than 5 μm due to its greater contact surface area, the predominance of the silt fraction in the sediment sampled confirms the high viability of OSA formation in the study area considered.

Figure 4.9 - Graphical representation of the percentage values of the sediment granulometric fractions along the Pardo River in the municipality of Canavieiras, south coast region of the state of Bahia, at the different sampling points

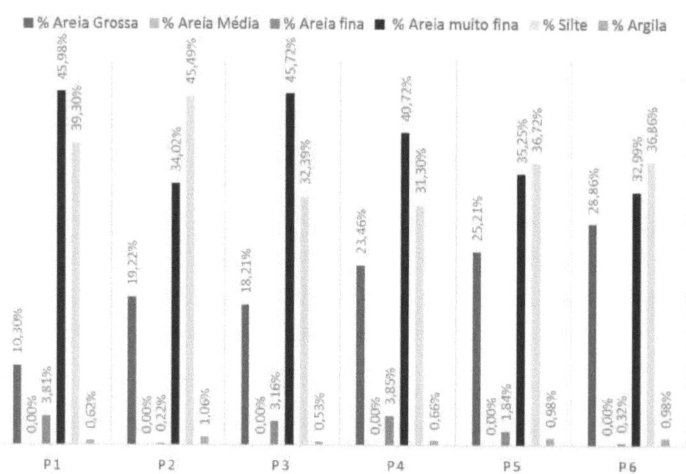

The crude oil from the Campos basin used for the tests had a density of 0.8823 g/mL and an average viscosity of 33.44 MPa·s. With regard to the distribution of hydrocarbons, the chromatogram of the Campos basin oil shows a higher composition of low molecular weight n-alkanes (nC10-nC17), suggesting typically marine oils (Figure 4.10). A Pristane/Phythane (Pr/Ph) ratio >1 indicates a depositional environment with oxidizing (oxic) conditions. A nC17/nC29 ratio >1 (nC29 is a terrestrial indicator and n-C17 is an indicator of marine algal material) indicates a predominantly marine input origin. Pr/nC17 and Ph/nC18 ratios lower than 1, and a CPI value of 1.07, are indicative of maturity for this oil (BARRAGAN, 2012). No evidence of biodegradation was observed for the characterized oil, justifying its use for the tests carried out.

Figure 4.10 - Total oil chromatogram for the Campos sedimentary basin sample

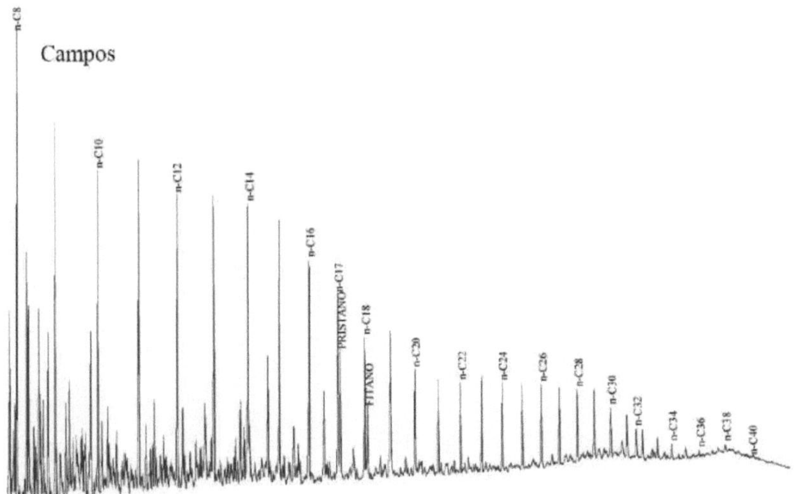

Source: BARRAGAN, 2012

4.3.2 Tests with reference substance Sodium Dodecyl Sulphate (DSS)

The acute toxicity tests with the reference substance Sodium Dodecyl Sulfate ($NaC_{12}H_{25}SO_4$) carried out simultaneously with the sensitivity tests in this study showed results within the acceptable limits described (< 9.25 mg/L) (CARIELLO, 2012). The $CL_{50\text{-}96h}$ values were 4.66, 5.73 and 4.38 mg/L, thus falling within the established sensitivity range.

Table 4.5 specifies the CL_{50} values found for each test, as well as the confidence intervals (95%) and the physico-chemical parameters measured.

Table 4.5 - CL values$_{50}$ with the confidence limits (95%) and representative physico-chemical parameters of the reference tests with DSS followed by the Mean, Standard Deviation and Coefficient of Variance

Reference test (DSS)	CL_{50} (mg/L)	Confidence Limit	pH	O.D. (mg/L)	Temperature (°C)
TR1	4,66	3,83 - 5,67	7,05	6,79	24,1
TR2	5,73	4,69 - 6,98	7,03	6,85	25,2
TR3	4,38	3,51 - 5,46	7,66	6,53	25,8
Average	4,92	**	7,25	6,72	25,03
Standard deviation	0,71	**	0,36	0,17	0,86
Coef. Vari.	0,14	**	0,05	0,03	0,03

Legend: TR = Reference Test; ** = Not applicable

4.3.3 Acute toxicological tests

The toxicity test with the aqueous phase of OSA formation was carried out between October 16 and 20 (96 hours), and the results of the analyses carried out and their adjustments using the Probit and Trimmed Spearman-Karber statistical methods are shown in Table 4.6, followed by the relative physico-chemical parameters recorded at the beginning and end of the analyses in Table 4.7.

Table 4.6 - Percentage mortality and survival data from the acute toxicity tests from the aqueous phase of the OSA formation simulation used to determine the LC_{50} including proportions adjusted for calculation and confidence limit (95%)

Concentration	Mortality	Survival	Adjusted proportions
0% (control)	0,33	9,67	0,0300
1%	2,00	8,00	0,1658
6,25%	2,33	7,67	0,2002
12,5%	2,33	7,67	0,2002
25%	2,67	7,33	0,2357
50%	4,33	5,67	0,4330
100%	6,00	4,00	0,6000
CL50 (%)		70,71	
Confidence Interval		33,9 - 151,9	

The toxicity test with the elutriate produced from the sediment matrix and homogenized oil was carried out between 29 October and 2 November (96 hours), and the results of the analyses carried out and their corrections are shown in Table 4.8 and Table 4.9, the physico-chemical parameters recorded at the beginning and end of the analyses.

Table 4.7 - Initial and final physico-chemical parameters representative of the acute toxicity test with the copepod *Nitokra* sp. from the aqueous phase of the OSA formation simulation, followed by the Mean, Standard Deviation and Coefficient of Variance.

PH			O.D. (mg/L)		Temperature (°C)		Conductivity (µS)	
Conc. (%)	Initial	Final	Initial	Final	Initial	Final	Initial	Final
0%	7,15	6,85	6,74	5,5	23,5	23,8	23,8	21,85
1%	7,10	7,05	6,6	5,78	23,8	24	20,49	21,97
6,25%	7,15	7,1	6,48	5,99	24	24,2	21,32	20,76

12,50%	7,15	7,2	6,37	5,57	24	23,6	20,95	23,21
25%	7,12	7,05	6,03	5,37	24,5	23,3	22,05	22,63
50%	7,15	6,75	5,54	5,68	24,6	23,3	20,15	22,49
100%	7,1	7,45	5,32	5,72	25,5	23	22,49	21,49
AVERAGE	7,13	7,11	5,815	5,585	24,65	23,3	21,41	22,455
DP	0,0244	0,2926	0,4743	0,1567	0,6245	0,2449	1,0606	0,7148
CV	0,003	0,0411	0,0815	0,0280	0,0253	0,0105	0,0495	0,0318

Legend: Conc. = Concentration; O.D. = Dissolved Oxygen.

It is important to note that, due to the response of the organisms tested in the pilot test, a percentage concentration of 0.5% was added, which comprises the intermediate value of mortality between the first percentage concentration tested and the control, helping the statistical fidelity of the analysis.

Table 4.8 - Percentage mortality and survival data from the acute toxicity tests on elutriate, used to determine the LC_{50} including proportions adjusted for calculation and confidence limit (95%)

Concentration	Mortality	Survival	Adjusted proportions
0% (control)	1,67	8,33	0,000
0,5%	2,67	7,33	0,0641
1%	2,67	7,33	0,0641
6,25%	5,33	4,67	0,4037
12,5%	5,67	4,33	0,4471
25%	7,00	3,00	0,6170
50%	10,00	0,00	1,0000
100%	10,00	0,00	1,0000
CL50 (%)		5,59	
Confidence Interval		5,59 - 2,87	

According to the data observed, the Lethal Concentration (LC) that affects 50% of the organisms of the species *Nitokra* sp. for exposure to elutriate is 5.59%, demonstrating high toxicity for the organism in question.

The strength of the linear associations between the test results and the physico-chemical parameters was checked using Pearson's linear correlation coefficient. The correlation coefficient is positive if the variables are directly related and negative if they are inversely related, with a maximum modular value of 1.

For the toxicological analysis in question, the values marked in red were significant for correlation considering p>.05. This analysis was used to verify the parameters that contribute most to toxicity, and strong positive correlations were found between percentage dilution and mortality, and inversely for dissolved oxygen (Table 4.10). The negative correlation for dissolved oxygen can be explained by the lower contribution of this factor to the increase in toxicity, since aeration is not a requirement in the cultivation of these organisms and, because they use oxygen in their respiratory processes, it can cause a reduction in its concentration in the medium and, depending on the magnitude, the consequent increase in toxicity (METCALF; EDDY, 2003).

Table 4.9 - Initial and final physico-chemical parameters representative of the acute toxicity test with the copepod *Nitokra* sp. from elutriate, followed by the Mean, Standard Deviation and Coefficient of Variance

PH			O.D. (mg/L)		Temperature (°C)		Conductivity (µS)	
Conc. (%)	Initial	Final	Initial	Final	Initial	Final	Initial	Final
0%	5,95	6,15	4,9	4,53	25,1	23,9	32,66	23,95
0,5%	5,95	5,65	5,25	5,88	24,3	22,2	29,82	29,26
1%	6,05	6,05	5,65	4,77	24,7	24,5	27,24	29,76
6,25%	6,05	5,9	4,72	4,53	24,5	24,1	24,98	23,18
12,50%	5,85	5,55	4,31	4,65	25,7	24,5	23,8	24,28
25%	5,95	5,45	4,8	4,19	25,4	25,5	23,4	21,22
50%	7,15	6,75	5,54	5,68	24,6	23,3	20,15	22,49
100%	7,1	7,45	5,32	5,72	25,5	23	22,49	21,49
AVERAGE	6,5125	6,3	4,9925	5,06	25,3	24,075	22,46	22,37
DP	0,7087	0,9678	0,5507	0,7627	0,4830	1,1500	1,6346	1,3856
CV	0,1088	0,1536	0,1103	0,1507	0,0191	0,0478	0,0728	0,0619

Legend: Conc. = Concentration; O.D. = Dissolved Oxygen.

Table 4.10 - Pearson's correlation matrix for physico-chemical parameters and toxicity tests with the copepod *Nitokra* sp.

	Conc.	Mort.	pH	O.D.	Temp.	Condt.
D.P.	1,00					

Mort.	0,94	1,00			
pH	0,47	0,47	1,00		
O.D.	-0,87	-0,81	-0,19	1,00	
Temp.	0,72	0,82	0,61	-0,42	1,00
Cdtv.	0,00	-0,29	0,03	-0,19	-0,52	1,00

Legend: Conc. = Concentration; Mort. = Mortality; O.D. = Dissolved Oxygen; Temp. = Temperature; Condt. = Conductivity.

Principal Component Analysis (PCA) was used to estimate the similarity of the data, representing it in two orthogonal dimensions. The two factors represented in Figure 4.11 explain the largest proportion of the total variance among all the linear combinations of the original data, with the second having a smaller proportion of total variance than the first, and so on.

Figure 4.11 - Graphical representation of the principal axes (1 and 2) of the principal component analysis of the correlation matrix (percentage dilutions, mortality and physico-chemical parameters)

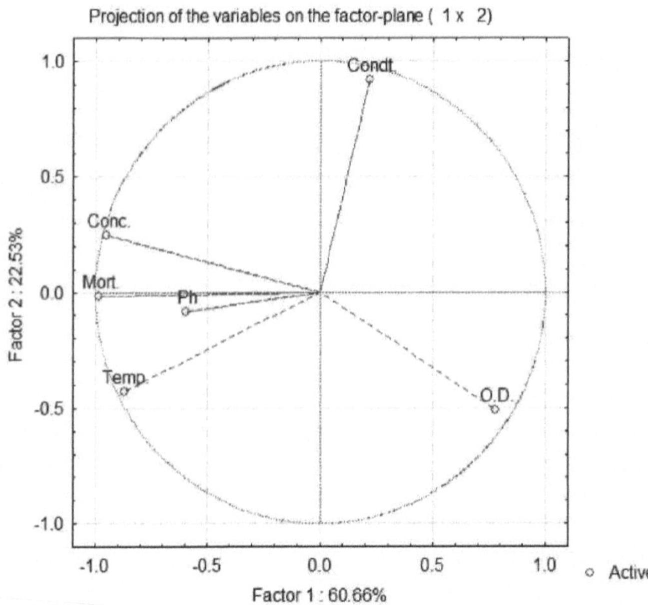

Principal component analysis allowed the data to be represented in a bivariate area (Figure 4.11). The two main axes - Factor 1 and 2, explain 83.19% of the total variance (22.53% and 60.66%).

In this graphical representation it is possible to corroborate the negative relationship between dissolved oxygen and the other parameters studied, verified through Pearson's correlation, and the strong positive correlation between the toxicity factors (concentration and mortality) with pH being the least influential among the physico-chemical parameters monitored.

With regard to the test's explainability factors, the *eigenvalues* represent the ratio of the variation between the groups to the variation within them, considering that the further away from 1, the greater the variation between the groups explained by the discriminant function.

Table 4.11 - *Eigenvalues*, total percentage of variance and cumulative variance for the explainability factors found for the correlation matrix

	Eigenvalues	% total variance	Cumulative eigenvalue	Cumulative %
Component 1	3,639330	60,65551	3,639330	60,6555
Component 2	1,351569	22,52615	4,990899	83,1817
Component 3	0,863636	14,39394	5,854536	97,5756
Component 4	0,098554	1,64257	5,953090	99,2182
Component 5	0,046837	0,78061	5,999927	99,9988
Component 6	0,000073	0,00122	6,000000	100,0000

Table 4.11 shows the six main factors that contribute to the total variance between the parameters, with the first component effectively having the greatest inter-group separating power.

The significant negative correlation between O.D. and toxicity factors (concentration and mortality) can be explained by the fact that dissolved oxygen is depleted mainly due to the decomposition of organic matter and the respiration of organisms, although some compounds such as ammonia increase their toxicity at low oxygen concentrations, resulting in compensatory physiological mechanisms to minimize the effects of hypoxia on these organisms (AMARAL, 2012).

According to Jiang and collaborators (2010), the oil fraction accommodated in the water shows high toxicity to marine zooplankton, especially in the first stages of life, since compounds such as HPA and benzene can accumulate in the tissues due to their lipophilic abilities. In some species, a tendency towards tolerance can manifest itself at different life stages and resistance to toxicity to oil compounds can be varied.

It is important to note that, even considering exposure to the OSA formation simulation experiment, where a low LC value was observed$_{50}$ (5.59), demonstrating a less toxic lethal potential, several studies have shown the effects of exposure to low concentrations for zooplankton, such as behavioral abnormalities in feeding (JENSEN et al., 2008; CALBET et al., 2007; BARATA et al., 2002), increased oxygen consumption (SMITH and HARGREAVES, 1984), abnormal growth and development, and decreased growth and reproduction rates (BELLAS and THOR, 2007;

BEJARANO et al., 2006; OLMSTEAD and LEBLANC, 2005).

4.4 CONCLUSIONS

The aim of this study was to evaluate the potential toxicity of the formation of OSA through microscale simulation experiments using the benthic copepod Nitokra sp. as a test organism, comparing it with the toxicity found for the elutriate formed from the percentage dilution of oil and sediment.

It was observed that, in spite of its different intrinsic characteristics, the concentrated aqueous portion of the simulation proved to be considerably less toxic (CL50 70.71%) than the elutriate formed from the percentage dilutions of the sediment with oil (CL50 5.59%).

With regard to the explanatory power of the influence of the monitored variables, the strong and positive correlations between the toxicity factors (concentration and mortality) stand out, to the detriment of dissolved oxygen, which showed a strong and negative correlation, while pH seemed to be the least determining factor for the outcome of the tests among the physico-chemical variables evaluated in this study.

Given the complexity and variability of hydrocarbons and water-insoluble compounds, it is necessary to emphasize the importance of relating toxicity analyses to the monitoring of physico-chemical parameters, since the confidence limits of the effects are not very small.

REFERENCES

AMARAL, K. G. C. **Correlation between toxicity factor and physico-chemical parameters for treated domestic effluents**. 2012. 105f. Dissertation (Master's Degree in Environmental Science and Technology) - Federal Technological University of Paranâ. Curitiba, 2012.

ARAÙJO-CASTRO, C. M. V. **Standardization and application of the marine copepod *Tisbe biminiensis* as a test organism in toxicological evaluations of estuarine sediments**. 2008. 104 f. Thesis (Doctorate in Oceanography) - Federal University of Pernambuco. Pernambuco, 2008.

BARATA, C.; BAIRD, D.; MEDINA. Determining the ecotoxicological mode of action of toxic chemicals in meiobenthic marine organisms: stage-specific short tests with *Tisbe battagliai*. **Marine Ecology Progress Series**, v. 230, p. 183-194, 2002.

BARRAGAN, O. L. V. **Geochemical characterization of oils from Latin America.** 2012. 109 f. Dissertation (Master's Degree in Geochemistry: Petroleum and Environment) - Postgraduate Program in Geochemistry, Federal University of Bahia. Salvador, 2012.

BAUMGARTEN, M. G. Z.; ROCHA, J. M. B; NIENCHESKI, L. F. H. **Manual of analysis in chemical oceanography.** Rio Grande: Editora da FURG, 1996. 132 p.

BEJARANO, A.C.; CHANDLER, G.T.; HE, L. Individual to population level effects of South Louisiana crude oil water accommodated hydrocarbon fraction (WAF) on a marine meiobenthic copepod. **Journal of Experimental Marine Biology and Ecology**, v. 332, n. 1, p. 49-59, 2006.

BELLAS, J.; THOR, P. Effects of selected PAHs on reproduction and survival of the calanoid copepod *Acartia tonsa*. **Ecotoxicology**, v. 16, n. 6 p. 465-474, 2007.

BRAILE, P. M.; CAVALCANTI, J. E. W. A. **Manual de tratamento de águas residuârias industriais.** Sao Paulo: CETESB, 1993. 764p.

BRITO, E. M.S.; DURAN, R.; GUYONEAUD, R.; GONI-URRIZA, M.; OTEYZA, T. G.; CRAPEZ, M. A. C.; ALELUIA, I.; WASSERMAN, J. C.A. A case study of *in situ* oil contamination in a mangrove swamp (Rio De Janeiro, Brazil). **Marine Pollution Bulletin**. v. 58, n. 3, p. 418-423, 2009.

CALBET, A.; SAIZ, E.; BARATA, C. Lethal and sublethal effects of naphthalene and 1, 2-dimethylnaphthalene on the marine copepod *Paracartia grani*. **Marine Biology,** v. 151, n. 1, p. 195-204, 2007.

CARIELLO, M. S. **Effect of naphthalene on the marine microalga *Dunaliella tertiolecta, the* sea urchin *Lytechinus variegatus,* and the estuarine microcrustaceans *Nitokra* sp. and *Leptocheirus plumulosus*.** 2012. 135 f. Dissertation (Master of Science), University of Sao Paulo. Sao Paulo, 2012.

EMBRAPA. National Soil Research Center. **Manual of chemical analysis of soils, plants and fertilizers.** SILVA, F. C. Campinas: Embrapa Informàtica Agropecuària; Rio de Janeiro: Embrapa Solos, 2009. 370 p.

FERRAZ, M. A.; ZARONI, L. P.; BERGMANN FILHO, T.U.; GASPARRO, M. R. SOUSA, E. C. P. M. **Egg hatching rate of the copepod *Nitokra* sp. in different salinities.** In: BRAZILIAN CONGRESS OF ECOTOXICOLOGY, 11, 2010. Bombinhas, Santa Catarina. **Proceedings...** Bombinhas, Santa Catarina. 2010.

FIORUCCI, A. R. The importance of dissolved oxygen in aquatic ecosystems. **Quimica Nova na Escola,** v. 22, p. 10 - 16, 2005.

FURLEY, T. H.; VITALI, M. M.; SOBREIRA, R. G.; ZARONI, L. P. **Acute sensitivity of the copepod *Nitokra* sp. to zinc sulfate at different salinities.** In: BRAZILIAN CONGRESS OF ECOTOXICOLOGY, 11, 2010. Bombinhas, Santa Catarina. **Proceedings...** Bombinhas, Santa Catarina. 2010.

GRASSHOFF, K.; KREMLING, K.; EHRHARDT, M. **Methods of seawater analysis** 3ed.

Florida: Verlage Chemie, 1999. 417p.

ITOPF. **Fate of Marine Oil Spills**: Technical Information Paper. London, UK. n 2, 2011. Available at:<http://www.itopf.com/information-services/publications/documents/

tip2fateofmarineoilspills.pdf >. Accessed on Dec. 10, 2013.

JENSEN, M. H.; NIELSEN, T.G.; DAHLL0F, I. Effects of pyrene on grazing and reproduction of *Calanus finmarchicus* and *Calanus glacialis* from Disko Bay, West Greenland. **Aquat. Toxicol.** v. 87, n. 2, p. 99-107, 2008.

JIANG, Z.; HUANG, Y.; XU, S.; LIAO, Y.; SHOU, L.; LIU, J.; CHEN, Q.; ZENG, J. Advance in the toxic effects of petroleum water accommodated fraction on marine plankton. **Acta Ecol. Sin**. v. 30, p. 8-15, 2010.

KHELIFA, A. B.; FIELDHOUSE, Z.; WANG, C.; YANG, M. LANDRIAULT, M.F. FINGAS, C.E. BROWN, AND L. GAMBLE. **A Laboratory Study on Formation of Oil- SPM Aggregates using the NIST Standard Reference Material 1941b**. In: PROCEEDINGS OF THE THIRTIETH ARCTIC AND MARINE OIL SPILL PROGRAM TECHNICAL SEMINAR, Environment Canada, Ottawa, Ontario, v. 1, p. 35-48, 2007.

KRUMBEIN, William Christian; SLOSS, Laurence Louis. Stratigraphy and sedimentation. **Soil Science**, v. 71, n. 5, p. 401, 1951.

LE FLOCH, S.; GUYOMARCH, J.; MERLIN, F.X.; STOFYN-EGLI, P.; DIXON, J.; LEE, K. The influence of salinity on oil-mineral aggregate formation. **Spill science and technology bulletin**, v. 8, n. 1, p.65-71, 2002.

METCALF; EDDY, INC. **Wastewater engineering**: treatment and reuse. 4 ed. Boston: McGraw-Hill, 2003.

NASCIMENTO, I. A.; SOUSA, E. C. P. M.; NIPPER, M. **Métodos em ecotoxicologia marinha:** aplicações para o Brasil. Sao Paulo: Artes Gràficas e Ind. Ltda, 2002. 262 p.

OLMSTEAD, A.W.; LEBLANC, G.A. Joint action of polycyclic aromatic hydrocarbons: predictive modeling of sublethal toxicity. **Aquat. Toxicol.** v. 75, n. 3, p. 253-262, 2005.

PINA, P.; BRAUNSCHWEIG, F.; SARAIVA, S.; SANTOS, M.; MARTINS, F.; NEVES, R. The Importance of Physical Processes in the Control of Eutrophication in Estuaries. **Sapientia.** Algarve, 2003. Available at < http://hdl.handle.net/10400.V40 >. Accessed on: 04 Apr. 2012.

QUEIROZ, A. F. de S.; CELINO, J. J. Mangroves and estuarine ecosystems in Todos os Santos Bay. In: QUEIROZ, A. F. DE S.; CELINO, J. J. **Avaliação de ambientes na Baia de Todos os Santos:** aspectos geoquimicos, geofisicos e biológicos. Salvador - BA: EDUFBA, 2008. p. 39-58.

SMITH, R. L.; HARGREAVES, B. R.; Oxygen consumption in *Neomysis Americana* (Crustacea:Mysidacea) and the effects of naphthalene exposure, **Mar. Biol.** v. 79, n. 2, p. 109-116, 1984.

STOFFYN-EGLI, P., LEE, K. Formation and characterization of oil-mineral aggregates. **Spill science and technology bulletin**, v. 8, n. 1, p. 31 - 44, 2002.

SUN, J.; ZHENG, X. A review of oil suspended particulate matter aggregation a natural process of cleansing spilled oil in the aquatic environment. **Journal of environmental monitoring: JEM**, v. 11, n. 10, p. 1801-1809. 2009.

UNITED STATES ENVIRONMENTAL PROTECTION AGENCY - USEPA. **Testing Manual EPA-821-R-02-012 - Methods for measuring the acute toxicity of effluents and receiving waters to freshwater and marine organisms.** Oct. 2002. Available at: < http://water.epa.gov/scitech/methods/cwa/wet/upload/2007_07_10_methods_wet_disk2_atx1- 6.pdf '. Accessed on 10 Jan 2014.

UNITED STATES ENVIRONMENTAL PROTECTION AGENCY - USEPA. **Testing Manual EPA-823-B-01-002 - Methods for collection, storage, and manipulation of sediments for chemical and toxicological analyses.** Oct 2001. Available at: < http://water.epa.gov/polwaste/sediments/cs/upload/collectionmanual.pdf '. Accessed on 04 Feb 2014.

WALKEY-BLACK, A. A critical examination of a rapid method for determining organic carbon in soils: Effect of variations in digestion conditions and of inorganic soil constituents. **Soil Science**, v. 63, p. 251- 263, 1947.

ZARONI, L. P.; BERGMANN FILHO, T.U.; FERRAZ, M. A.; GASPARRO, M. R. &SOUSA, E. C. P. M. **Egg hatching rate of the copepod *Nitokra* sp. in different types of sediment.** In:

BRAZILIAN CONGRESS OF ECOTOXICOLOGY, 11, 2010. Bombinhas, Santa Catarina. **Proceedings...** Bombinhas, Santa Catarina. 2010.

5 DETERMINING THE TOXIC POTENTIAL OF THE FORMATION OF OIL-SUSPENDED PARTICULATE MATERIAL (OSM) AGGREGATES IN A SIMULATION EXPERIMENT AT THE MICRO-SCALE LEVEL

Summary:

The aggregation of suspended particulate matter (SPM) and oil droplets in environments with a certain hydrodynamic energy can lead to the formation of Oil-SpM Aggregates (OSA). A laboratory simulation was carried out to investigate the possible toxic potential of OSA formation, using three different concentrations of particulate matter (50, 200, 300 mg/L) in a microscale experiment. For this purpose, we used mangrove sediment samples collected along the estuary of the Pardo River, in the municipality of Canavieiras, on the southern coast of the state of Bahia, and oil samples obtained from the Campos sedimentary basin. The procedure was carried out using acute exposure toxicology tests to determine the CL_{50} (50% lethal concentration) using the microcrustacean *Artemia salina* as the test organism. Serial dilutions were carried out using surface and bottom sampling in order to characterize the toxicity in different plots. The concentration with the highest toxic potential was 200 mg/L, with equal values for surface and bottom (CL_{50} 7.91%), while the concentration with the lowest toxic potential was 300 mg/L (CL_{50} 31.5%) for surface samples. There was a negative correlation between the oxidation-reduction potential and the hydrogen potential (only in samples with 200 mg/L of sediment), and a positive correlation between the toxicity factors (percentage dilution and mortality) and the other parameters monitored.

5.1 INTRODUCTION

The interaction between droplets of oil suspended in water and suspended particulate matter (SPM) has been the subject of research for several studies, not only on its properties, but also on the variables that characterize its formation (OWENS, 1999; KHELIFA et al, 2002 - 2005b; LEE et al., 1998 - 2003; MUSCHENHEIM; LEE, 2002; OMOTOSO, 2002; OWENS; LEE, 2003; PAYNE et al., 2003; AJIJOLAIYA, 2004). Due to its characteristics, OSA is an effective cleaning method for treating oil in water. Oil droplets are easily fragmented into sizes of a few micrometers, which leads to the rapid transfer of oil stains from the sea surface to the water column, increasing its contact surface (LE FLOCH et al., 2002; STOFFYN-EGLI; LEE, 2002). The formation of OSA in the environment through oil-MPS interaction also increases the rate of dispersion and biodegradation of the oil, as well as preventing it from adhering to the sediment (LEE; STOFFYN-EGLI, 2001; OWENS, 1999).

It is important to consider that the disposal of oil in droplets can increase the concentration of hydrocarbons in solution, also increasing the bioavailability of these products, promoting not only an increase in natural degradation but also an increase in the toxicity of the oil due to exposure to compounds such as benzene, toluene and xylene. These compounds have considerable solubility in water compared to others, which makes marine organisms more vulnerable, since they absorb these contaminants through their tissues, gills, direct ingestion of water or contaminated food (ITOPF, 2013).

Acute toxicity tests are widely used to measure the possible effects of toxic substances on certain species over a short period of time compared to the life span of the test organism in question. It assesses the dose or concentration of a toxic agent that would produce a specific measurable reaction in a test organism over a short period of time, usually 24 to 96 hours. The main effect measured in acute toxicity studies is lethality or some other behavior that precedes it, such as immobility. Acute toxicity tests allow for EC50 (Effective Concentration) and LC50 (Lethal Concentration) values. The values of effective and lethal concentrations are always related to 50% of the organisms, because these responses are more reproducible and can be estimated with a greater degree of reliability and are more significant for extrapolation to a population (COSTA et al., 2008).

The microcrustacean *Artemia salina* (LEACH, 1819) is used extensively as a test organism in ecotoxicological laboratory tests, mainly related to estuarine, marine and hypersaline environments. Among the intrinsic characteristics that make this organism widely used in laboratory tests are the wide range of tolerance to salinity (5-250) and temperature (6-25°C); short life cycle; high adaptability to environmental adversities and test and cultivation conditions; high fecundity (bisexual or parthenogenetic reproduction); small size; non-selective diet and high adaptability to various nutrients, ensuring reliability, viability and cost-effectiveness of the tests (NUNES et al., 2006).

The main aim of this research is to evaluate the sensitivity of the microcrustacean *Artemia salina* in toxicological tests with static tests without renewal where it is possible to determine the adverse effects of exposure to samples taken in tests where the formation of OSA was simulated in a microscale experimental protocol, from serial dilutions of the surface and bottom fractions, evaluating the effects on the survival (acute toxicological test) of this organism in these different plots.

5.2 MATERIALS AND METHODS

In this study, acute toxicological tests were carried out in experiments simulating the formation of Oil-Suspended Particulate Matter Aggregates on a micro-scale (pilot scale). The acute tests were carried out in serial percentage dilutions of surface and bottom samples.

In parallel with each test, tests were carried out with a reference substance (Sodium Dodecyl Sulphate - DSS), to draw up a control chart with the aim of illustrating the regularization of the sensitivity of the cultivation of organisms used for the tests. Each test was accompanied by analyses of the associated physico-chemical parameters (pH, dissolved oxygen, temperature and conductivity), at the beginning and end of the tests, in order to control the basic conditions of exposure and support the interpretation of the results.

The data produced by the acute tests with the microscale simulations and with the reference substances were expressed as mortality and survival data for the adult individuals (CL_{50}) using the mathematical Probit method or Trimmed Spearman-Karber Method, as recommended by the USEPA (2002), associating them with 95% confidence intervals and producing responses related to mortality (or survival), which will be given for each test in comparison with the control group.

5.2.1 Sediment collection and storage

Sediment was sampled at six points during low tide along the Rio Pardo, near the municipality of Canavieiras, on the southern coast of the state of Bahia, taking care to collect the sediment in the first five centimeters with a stainless steel spatula that had been previously decontaminated and treated with local water. The sediment was collected within the limits of a transect approximately four meters long, where 20 portions of the mangrove substrate were randomly collected, homogenized in a glass tray and properly stored at a temperature of up to 4 °C. The sediment's physical and chemical parameters (pH, redox potential, temperature, salinity, conductivity, dissolved oxygen, total solids and turbidity) were measured *in situ* using a multi-parameter probe.

To carry out the simulations, each sediment sample was calcined in a muffle furnace at 450 °C for 6 hours, homogenized, sieved through a 100 μm mesh and kept in a refrigerator at 4 °C in glass containers for a period of 24 hours, until the sediment was distributed in the test containers for the bioassays to begin.

5.2.2 Oil and sediment characterization

The characterization of the oil and sediment matrices was carried out at the Petroleum Studies Laboratory (LEPETRO/IGEO/UFBA), and included analyses of sediment granulometry, API grade (density) and viscosity/fluidity, HTP and HPA in the oil. Sediment analysis did not exceed 48 hours after collection.

In the laboratory, after freeze-drying, sieving (2 mm) and homogenizing the samples, particle size analyses were carried out using a Cilas 1064 laser diffraction particle analyser, with the sample pre-treated according to Embrapa (2009).

The simulations used crude oil from the Campos basin, whose density was measured six times with a densitometer (DMA 5000 Density Meter) at 15°C, and its viscosity was also determined three times under the same temperature conditions using a rotational viscometer (Haake Viscotester VT500).

The distribution of hydrocarbons was also used to characterize the oil by gas chromatography coupled to a Varian CP 3800 flame ionization detector (GC-FID) equipped with a 60 m long, 0.25 mm internal diameter DB5 capillary column, a 0.25 μm thick stationary phase and a flow of carrier gas (helium).

5.2.3 Cultivation and acquisition of organisms

The *Artemia salina* cysts were purchased from an aquarium and kept away from light in silica gel desiccators to ensure they were free from contaminants and to preserve their hatching rate.

They were then hatched in glass aquariums with a capacity of 1L with constant aeration, following the methodology suggested by Veiga and Vital (2002), not exceeding a concentration of 0.5 g/L of cysts. A salinity of 15 and an average temperature of 25°C were maintained in each tank.

48 hours after the cysts had hatched, the organisms were collected using a glass pasteur pipette to carry out the reference tests and toxicity tests, since at this stage they are suitable for testing because they are already beginning to filter, making them more sensitive and reducing the variability of the test (SORGELOOS et al., 1978).

5.2.4 Tests with reference substance Sodium Dodecyl Sulphate (DSS)

The stock solution was prepared in 100 mL flasks in which 10 mg of sodium dodecyl sulphate was added for every 100 mL of salt water at salinity 15, which is obtained by diluting 30g of artificial sea salt in 500 mL of water, and volumized to one liter.

Three reference tests were carried out with the substance Sodium Dodecyl Sulphate to assess the sensitivity range of the organisms used in the test and to draw up the toxicity control chart. The samples were diluted in volumetric flasks to concentrations of 0, 9, 12, 16, 21, 27 and 35 mg/L.

Each dilution unit was carried out in triplicate with glass test vials containing 10 mL of the solution and ten organisms. The containers were placed on a tray and covered to minimize evaporation losses.

The conditions for the test were maintained respecting the photoperiod of 12h light and 12h dark, temperature of 25°C for 48h. The *endpoint* observed is the CL_{50} of 48 hours. In other words, after a two-day exposure, the living and dead organisms in each test are counted using a microscope or magnifying glass.

5.2.5 Microscale simulation of oil-material aggregate formation

Suspended Particulates (OSA)

The protocol consists of simulating the environmental conditions that determine the formation of OSA, starting with erlenmeyers with 250 mL of artificial brackish water at salinity 15, and weighing fractions of sediment for each one, in order to respect the concentration value of 0.05; 0.3 and 0.3 g/L of sediment.

The sediment was transferred to erlenmeyer flasks with the help of a pisser filled with artificial brackish water, and each one was covered with aluminum foil, placed on a shaking table for one minute and left to rest overnight at a temperature of 15°C in a darkened room.

On the following day, each of the conical flasks spent another minute on a shaker table at 2.1 Hz (126 RPM) and then 0.05g (50 mg or 0.05 cm^3) of oil was added, taking the expected value of 50 mg. This was followed by a further three hours of stirring on a reciprocating shaker table at 126 RPM and another overnight rest. The samples were used as soon as the preparation was finished, and fractions were randomly collected from the surface and bottom.

5.2.6 Acute toxicological tests

For the acute toxicity tests of the surface and bottom samples from the OSA formation simulation experiments, artificial brackish water at salinity 15 was used to dilute the fractions collected from the surface using an automatic pipette in the proportions described in Table 5.1.

For each test, concentrations of 0.05; 0.02 and 0.03 mg/L of sediment were used for approximately 0.05g of oil, four percentage serial dilutions (100%, 50%, 25%, 12.5%) and a positive control (blank 0%) in the three sediment concentrations, evaluating surface and bottom. Ten adult males or non-ovulated females were added to each container containing 15 mL of the dilution solution made from the elutriate.

Table 5.1 - Values used for surface and bottom fractions for each simulation of OSA formation and artificial brackish water for acute toxicity tests

Percentage Dilutions	Aqueous phase (mL)	Artificial brackish water (mL)
100%	10	-
50%	5	5
25%	2,5	7,5
12,5%	1,25	8,75

The salinity of the water was monitored using a portable refractometer and the physico-chemical parameters of all the tests were measured using a portable pH meter with an accuracy of ±0.01 pH units; O.D. with a microprocessor meter.

with an accuracy of ±0.05%; temperature with a thermometer attached to the oximeter with an accuracy of ±0.05 °C and conductivity with a conductivity meter, digital handheld with an accuracy of ±0.05%. The conditions for the acute toxicity test are described in Table 5.1.

Table 5.1 - Conditions for the acute toxicity test (mortality) with *Artemia salina* for samples from the surface and bottom fractions of the OSA formation simulation experiments in a mangrove environment

Parameters	Conditions
Temperature Salinity Photoperiod Test system Test duration Volume of test solution Replicates per dilution Organisms per replicate Age of organisms Effect observed Validity of test	25±1°C 15 Natural (12h light/12h dark) Static 48h 10 mL 3 10 Adults Lethality (CL50) Minimum 70% survival in control

5.3 RESULTS AND DISCUSSION

5.2.7 Oil and sediment characterization

The results of the characterization of the sediment and oil used in this study will be described here, taking into account the field and sediment collection stage, which includes the measurement of physico-chemical parameters, as well as the laboratory analyses carried out. These results demonstrate the need to characterize the sediment used in the tests in order to control the experimental conditions evaluated. Table 5.2 shows the geographic coordinates and physico-chemical parameters of the collection points in the mangrove substrate of the study region.

The coordinates are directly related to the sampling points, as can be seen in Table 5.2, associated with the collection times described therein, which obeyed the restriction of collection at low tide (around 0.3m). The values for pH, Eh, temperature, salinity, conductivity, dissolved oxygen, total solids and turbidity are shown graphically below.

Table 5.2 - Geographical coordinates and descriptive physico-chemical parameters at the collection points on the banks of the Pardo river, municipality of Canavieiras, south coast region of the state of Bahia, at the different sampling points

	Coordinates	Time	pH	Eh (mV)	Temp. (°C)	Salt.	Condt. (S/m)	O.D. (mL/L)	S.T.D. (mg/L)	Turbidity (NTU)
				Physico-chemical parameters (*in situ*)						
P1	S 15° 41' 58" W 38° 55' 54"	07:35	9,91	45	26,56	30	30,8	15,2	17,4	121,0
P2	S 15° 41' 36" W 38° 56' 21"	08:57	8,55	163	29,62	20	35,7	18,3	21,8	26,6
P3	S 15° 40' 52" W 38° 56' 24"	09:22	5,93	325	30,02	12	23,4	17,66	14,5	30,6
P4	S 15° 41' 26" W 38° 57' 05"	10:03	7,89	245	30,82	6	15,5	21,04	9,61	49,1
P5	S 15° 41' 25" W 38° 57' 35"	10:56	7,61	240	30,37	6	14,8	19,63	9,16	99,2
P6	S 15° 40' 58" W 38° 58' 25"	11:47	7,91	222	30,08	5	12,3	18,95	7,62	179

Legend: pH = Hydrogen Potential; Eh = Reduction Potential; Temp. = Temperature; Sal. = Salinity; Condt. = Conductivity; O.D. = Dissolved Oxygen; S.T.D. = Total Dissolved Solids; NTU = Nephelometric Turbidity Units.

Figure 5.1 - Graphical representation of pH values along the Rio Pardo in the municipality of Canavieiras, south coast region of the state of Bahia, at the different sampling points

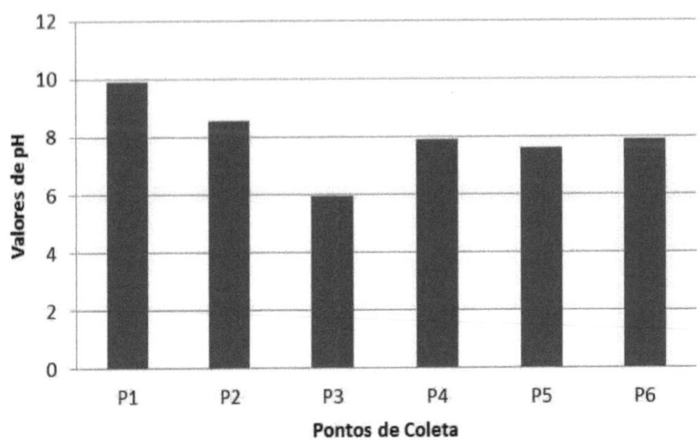

The pH values were neutral to slightly basic (with the exception of point 3, which was relatively

57

acidic), with little variation between the six sampling points, with a maximum recorded at P1 (9.91) and a minimum recorded at P3 (5.93), which may be indicative of the uniformity of characteristics between these sites, as a result of the similar contributions of marine and fresh waters at the sampling points (Figure 5.1).

With regard to Eh values, there was a wide variation between the maximums and minimums observed at points 3 and 1 (325.0 mV) and 5 (45.0 mV), respectively, all of which show a significant oxidizing characteristic, which represents the instability of this environment during the measurement of this parameter (Figure 5.2).

Figure 5.2 - Graphical representation of Eh values (in mV) along the Pardo River in the municipality of Canavieiras, southern coastal region of the state of Bahia, at the different sampling points

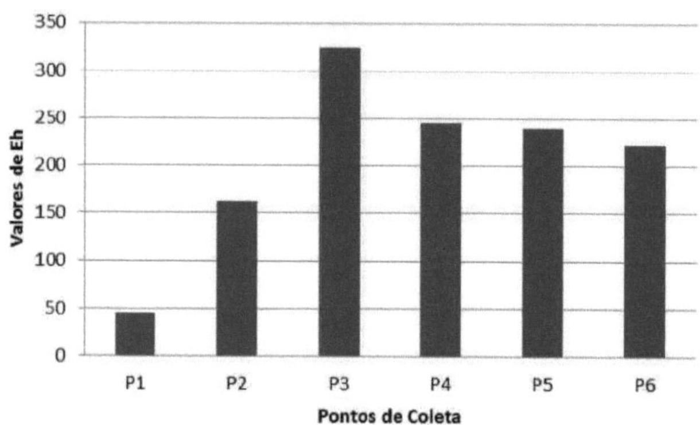

The temperature variation did not exceed a difference of 4.2 °C between the points, with the maximum value recorded at P4 (30.8 °C) and the minimum value seen at P1 (26.5 °C). This variation in temperature may be related to the influence of climatic conditions depending on the time and place of collection, as well as the intrinsic characteristics of the points sampled (Figure 5.3).

Figure 5.3 - Graphical representation of temperature values (in °C) along the Rio Pardo in the municipality of Canavieiras, south coast region of the state of Bahia, at the different sampling points

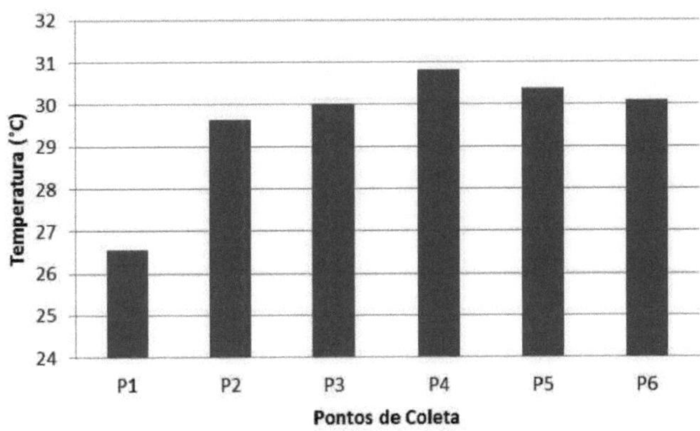

With regard to salinity, as expected for an estuary environment, it remained relatively high at the first three points (30, 20, 12) and decreased significantly at the last three (6, 6 and 5, respectively). In this case, it is important to note that there is a gradual reduction in values, where the highest indices were found at sampling points that are influenced by salinity downstream and the lowest upstream of the river (Figure 5.4). This behavior can be explained by the penetration of the tide, which causes samples collected in mangroves that are closer to the mouth of the estuary to have higher salinities.

The parameters pH, Eh and salinity are temperature-dependent, increasing and decreasing in a non-linear way according to temperature variations. All of these factors, together or in isolation, contribute to the cations that are sorbed to the particles that make up the mangrove substrates becoming bioavailable (QUEIROZ; CELINO, 2008). Therefore, the results of the physico-chemical parameters at the different sampling stations are crucial for correlating their possible effects as influences on the toxicity relationship found in the sediments.

Figure 5.4 - Graphical representation of salinity values along the Rio Pardo in the municipality of Canavieiras, southern coastal region of the state of Bahia, at the different sampling points

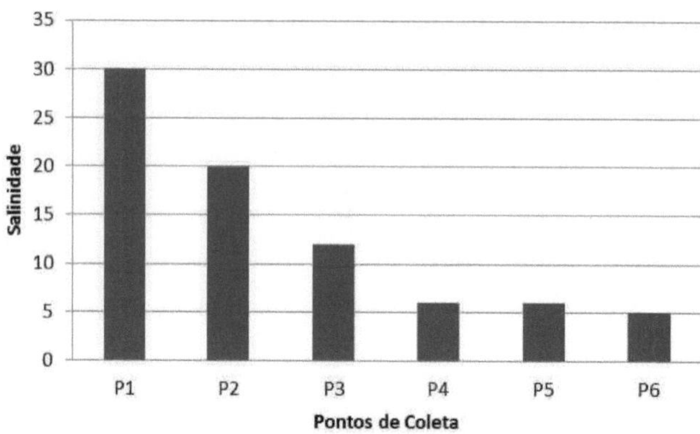

Dissolved oxygen values ranged from a maximum of 21.04 mL/L to a minimum of 15.20 mL/L (Figure 5.5). This variable is influenced by the composition of the sediments, the temperature, the salinity of the water and hydrodynamic factors. It offers indications of the natural conditions of the environment and makes it possible to detect possible environmental impacts such as eutrophication and organic pollution. Carbon dioxide, molecular oxygen, nitrite and nitrate ions and the water itself are the main sources of dissolved oxygen (BAUMGARTEN et.al., 1996).

It can be seen that dissolved oxygen had a behavioral pattern of values very similar to temperature. The temperature values at the different points can be considered a determining factor for the distribution of dissolved oxygen values, since temperature determines the solubility of gases (FIORUCCI, 2005).

With regard to the values found for conductivity (Figure 5.6), a wide variation was observed, with the highlight being the maximum value found in P2 (35.7 mS). It is possible to observe an inverse relationship between the behavior of e compared to conductivity, which can be interpreted from the perspective of the same criterion since the presence of dissolved ions can have a huge influence on the electrical conductivity values recorded (QUEIROZ; CELINO, 2008).

Figure 5.5 - Graphical representation of dissolved oxygen values (in mL/L) along the Rio Pardo in the municipality of Canavieiras, south coast region of the state of Bahia, at the different sampling points

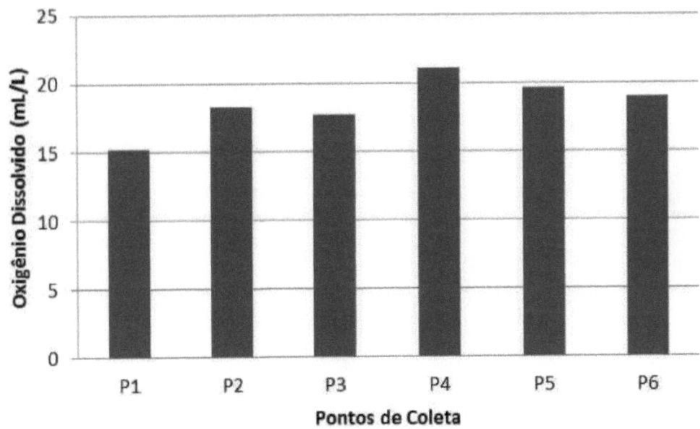

The highest value for total dissolved solids was found in P2 (21.8 mg/L) and the lowest in P6 (7.6 mg/L) (Figure 5.7), so the total solids values were directly proportional to those found for conductivity. Sampaio and colleagues (2007) had already carried out studies showing that there is a strong linear correlation between these two parameters, which can be explained by the fact that they are in solution and in a colloidal state (BRAILE; CAVALCANTI, 1993).

Figure 5.6 - Graphical representation of conductivity values (in S/m) along the Rio Pardo in the municipality of Canavieiras, southern coastal region of the state of Bahia, at the different sampling points

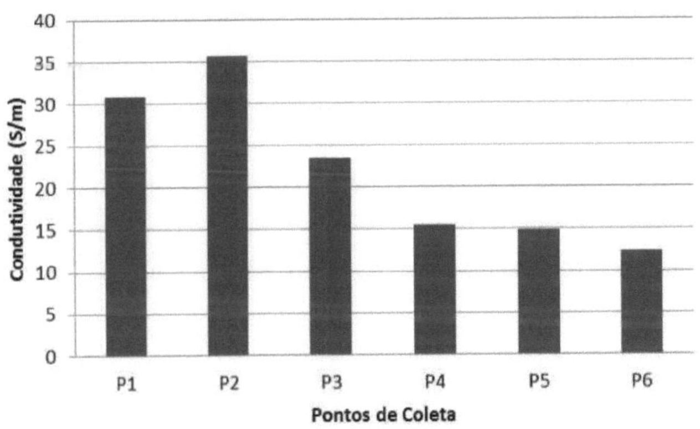

Figure 5.7 - Graphical representation of total dissolved solids values (in mg/L) along the Rio Pardo in the municipality of Canavieiras, south coast region of the state of Bahia, at the different sampling points

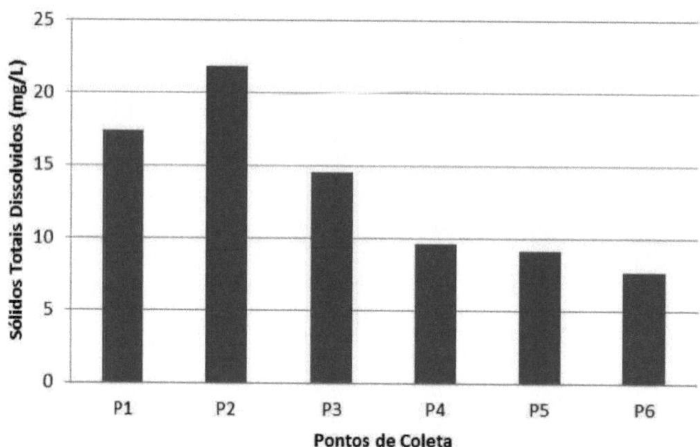

For turbidity, the highest value was found in P6 (179.0 NTU) followed by P1 (121.0 NTU) and the lowest value was recorded in P2 (20.6 NTU) (Figure 5.8). According to Pina et al. (2003) the high turbidity of the estuary is a consequence of the high tidal flow, the river load, the large expanse of intertidal areas and the high susceptibility to wave generation which intensifies resuspension.

Each of these values shows a set of environmental conditions in the study area that would enable and sometimes mitigate the formation of OSA in these locations, considering the factors that influence the formation of these aggregates in a natural environment.

Figure 5.8 - Graphical representation of turbidity values (in NTU) along the Rio Pardo in the municipality of Canavieiras, south coast region of the state of Bahia, at the different sampling points

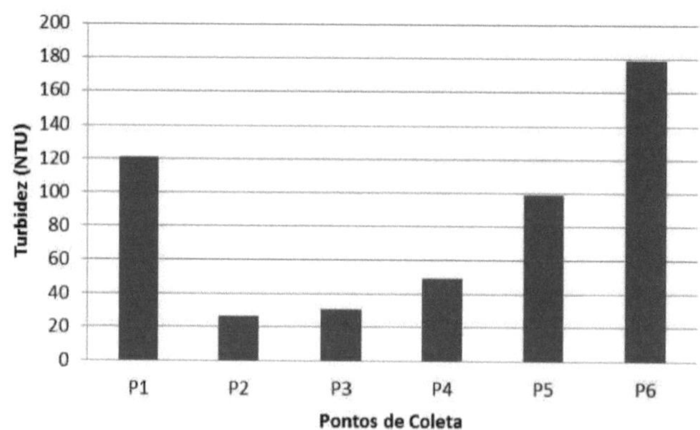

With regard to the granulometric profile of the collection points (Figure 5.9), the averages for the values found in the analyses show a predominance of the fractions considered to be very fine sand (62.5 - 125 µm) and silt (3.9 - 62.5 µm) according to the granulometric classification proposed by the Krumbein scale (1951). Considering that, according to Sun & Zeng (2009), granulometry is a determining factor in the formation and fate of the OSAs formed and that OSA formation is attenuated in sediment with a granulometry size of less than 5 µm due to its greater contact surface area, the predominance of the silt fraction in the sediment sampled confirms the high viability for OSA formation in the study area considered.

The crude oil from the Campos basin used for the tests had a density of 0.8823 g/mL and an average viscosity of 33.44 MPa·s. With regard to the distribution of hydrocarbons, the chromatogram of the Campos basin oil shows a higher composition of low molecular weight n-alkanes (nC10-nC17), suggesting typically marine oils (Figure 5.10). A Pristane/Phythane (Pr/Ph) ratio >1 indicates a depositional environment with oxidizing (oxic) conditions. A nC17/nC29 ratio >1 (nC29 is a terrestrial indicator and n-C17 is an indicator of marine algal material) indicates a predominantly marine input origin. Pr/nC17 and Ph/nC18 ratios lower than 1, and a CPI value of 1.07, are indicative of maturity for this oil (BARRAGAN, 2012). No evidence of biodegradation was observed for the characterized oil, justifying its use for the tests carried out.

Figure 5.9 - Graphical representation of the percentage values of the sediment granulometric fractions along the Pardo River in the municipality of Canavieiras, south coast region of the state of Bahia, at the different sampling points

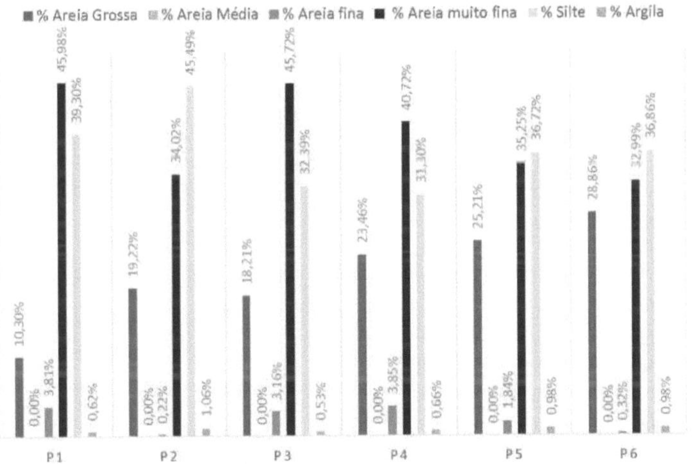

Figure 5.10 - Total oil chromatogram for the Campos sedimentary basin sample

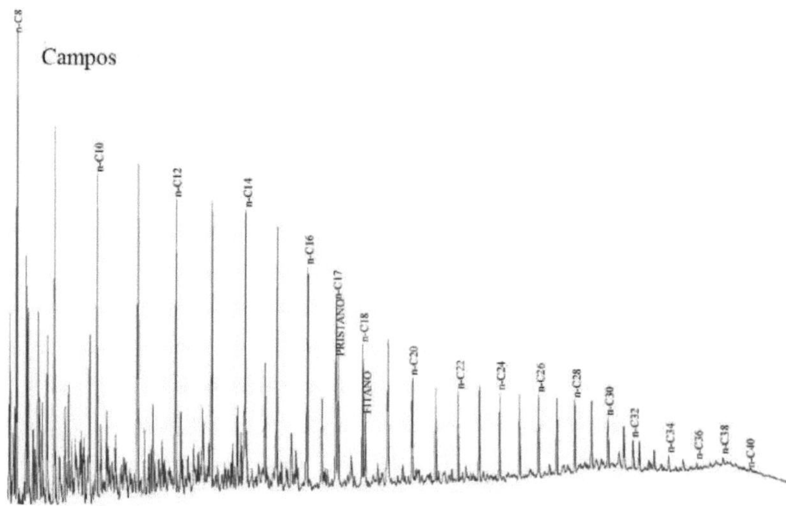

Source: BARRAGAN, 2012

5.3.1 Tests with reference substance Sodium Dodecyl Sulfate (DSS):

The acute toxicity tests with the reference substance Sodium Dodecyl Sulfate ($NaC12H25SO4$) carried out simultaneously with the sensitivity tests in this study showed results within the acceptable limits described (13 - 30.9 mg/L) (VEIGA; VITAL, 2002). The CL50-48h values were 27.06, 14.19 and 14.51 mg/L, thus falling within the established sensitivity range.

Table 5.3 specifies the CL_{50} values found for each test, as well as the confidence intervals (95%) and the physico-chemical parameters measured.

Table 5.3 - CL values$_{50}$ with the confidence limits (95%) and representative physico-chemical parameters of the reference tests with DSS followed by the Mean, Standard Deviation and Coefficient of Variance

Reference test (DSS)	CL_{50} (mg/L)	Confidence Limit	PH	O.D. (mg/L)	Temperature (°C)
TRI	27,06	23,18 - 31,58	7,12	6,57	25,1
TR2	14,19	12,5 - 16,1	7,33	6,65	24,8
TR3	14,51	11,75 - 18,17	7,06	6,93	25,3
Average	18,58	**	7,17	6,71	25,06
Standard deviation	7,33	**	0,14	0,18	0,25
Coef. Vari.	0,39	**	0,01	0,02	0,01

Legend: TR = Reference Test; ** = Not applicable

5.3.2 Acute toxicological tests:

The acute toxicity tests were carried out with the aqueous phase of the OSA formation simulation protocol (MOREIRA, 2014) and with elutriate produced from the homogenized sediment and oil matrix. Each of these tests was characterized according to the number of partial mortalities adjusted to produce CL50 values associated with a 95% confidence interval.

The toxicity test with the aqueous phase of OSA formation was carried out between February 22 and 24, 2014 (48 hours), and the results of the analyses carried out and their adjustments made using the Probit and Trimmed Spearman-Karber statistical methods are shown in Table 5.4, and later, in Table 5.5, the relative physico-chemical parameters recorded at the beginning and end of the analyses are presented.

Table 5.4 - Percentage mortality and survival data from the acute toxicity tests from the aqueous phase of the OSA formation simulation used to determine the LC_{50} including proportions adjusted for calculation and confidence limit (95%) at the different sediment concentrations (50, 200, 300 mg/L).

50 mg/L	Surface			Fund		
Percentage Dilution	Mortality	Survival	Adjusted proportions	Mortality	Survival	Adjusted proportions
1%	3,4	6,6	0,334	4,4	5,6	0,44
12,50%	3	7	0,3	4,4	5,6	0,44
25%	5,4	4,6	0,54	4,7	5,3	0,47
50%	7	3	0,7	5	5	0,5
100%	8,67	1,33	0,867	5,7	4,3	0,57
CL50 (%)	25			70,71		
Confidence Interval	13,01 -48,03			**		

200 mg/L	Surface			Fund		
Percentage Dilution	Mortality	Survival	Adjusted proportions	Mortality	Survival	Adjusted proportions
1%	4,7	5,3	0,47	4,4	5,6	0,44
12,50%	5,7	4,3	0,57	5,4	4,6	0,54
25%	6	4	0,6	6	4	0,6
50%	6,7	3,3	0,67	6,67	3,33	0,667
100%	7	3	0,7	6,67	3,	330,667
CL50 (%)	7,91			7,91		
Confidence Interval	0,31 -204,80			0,31 -204,80		

300 mg/L	Surface			Fund		
Percentage Dilution	Mortality	Survival	Adjusted proportions	Mortality	Survival	Adjusted proportions
1%	4	6	0,4	3,34	6,66	0,334

12,50%	4,4	5,6	0,44	5	5	0,5
25%	5	5	0,5	5	5	0,5
50%	5,4	4,6	0,54	6,67	3,33	0,667
100%	8	2	0,8	7,7	2,3	0,77
CL50 (%)		31,5			13,5	
Confidence Interval		7,10 - 139,68			2.28 -77.51	

Caption: ** = Not achieved.

Table 5.5 - Initial and final physico-chemical parameters (representative of the acute toxicity test with the microcrustacean *Artemia salina* from the aqueous phase of the OSA formation simulation, followed by the Mean, Standard Deviation and Coefficient of Variance at the different sediment concentrations)

50 mg/L	Surface								Fund							
	pH		O.D (mg/L)		Temperature (°C)		Eh (mV)		pH		O.D (mg/L)		Temperature (°C)		Eh (mV)	
Conc. (%)	Initial	Final	Initial	Final	Initial	Final	Initial	Final	Initial	Final	Initial	Final	Initial	Final	Initial	Final
12,50%	7,15	6,6	9,18	9,56	21,1	21,9	-24,6	12,8	7,62	6,68	9,23	9,52	21,5	22,8	-46,8	-2,3
25%	7,19	6,02	9,13	6,51	21,5	22,7	-27,6	19,8	7,64	6,15	9,23	9,4	21,4	22,5	-43,1	4,9
50%	6,84	6,55	9	9,41	23,7	23,1	-19,3	11,7	7,55	6,48	9,24	9,39	21,4	22,6	-37,4	-3,4
100%	7,07	8,45	9,14	9,63	22,2	22,9	-21,2	-88,7	7,27	6,26	9,26	9,44	21,5	21,3	-36,1	16,3
AVERAGE	7,0625	6,905	9,112	8,777	22,125	22,65	-23,15	-11,1	7,52	6,392	9,24	9,437	21,45	22,3	-40,8	3,875
DP	0,1564	1,062	0,078	1,514	1,1441	0,5259	3,675	51,857	0,171	0,235	0,0141	0,059	0,0577	0,6782	4,997	9,064
CV	0,0221	0,153	0,008	0,172	0,0517	0,0232	-0,158	-4,671	0,022	0,036	0,0015	0,006	0,0026	0,0304	-0,12	2,339

200 mg/L	Surface								Fund							
	pH		O.D (mg/L)		Temperature (°C)		Eh (mV)		pH		O.D (mg/L)		Temperature (°C)		Eh (mV)	
Conc. (%)	Initial	Final	Initial	Final	Initial	Final	Initial	Final	Initial	Final	Initial	Final	Initial	Final	Initial	Final
12,50%	7,61	6,6	9,25	9,04	21,9	20	-45,7	3,9	7,59	6,77	9,22	8,89	22	21,4	-46,4	5,9
25%	7,29	6,82	9,13	9,12	21,8	20,8	-46,3	0,8	7,52	6,58	9,21	8,92	21,8	21,2	-43,4	9,9
50%	7,6	7,07	9,25	9,16	21,9	24,9	-47,5	-22,6	7,54	6,61	9,18	8,91	21,9	22,8	-44,3	12,6
100%	7,6	7,04	9,27	9,33	21,6	24,4	-47,3	-15,7	7,57	6,36	9,22	9,09	22,1	24,1	-46,5	15,6
AVERAGE	7,525	6,882	9,225	9,162	21,8	22,525	-46,7	-8,4	7,555	6,58	9,207	8,952	21,95	22,375	-45,15	11
DP	0,1567	0,218	0,0640	0,122	0,1414	2,4837	0,8485	12,79	0,0310	0,168	0,018	0,092	0,1290	1,3524	1,545	4,120
CV	0,0208	0,031	0,0069	0,013	0,0064	0,1102	-0,018	-1,52	0,0041	0,025	0,002	0,010	0,0058	0,0604	-0,034	0,374

300 mg/L	Surface								Surface							
	pH		O.D (mg/L)		Temperature (°C)		Eh (mV)		pH		O.D (mg/L)		Temperature (°C)		Eh (mV)	

Conc. (%)	Initial	Final	Initial	Final	Initial	Final	Initial	Final	Initial	Final	Initial	Final	Initial	Final	Initial	Final
12,50%	7,79	6,92	9,24	8,73	21,7	25	-52,2	-4,3	7,77	7,04	9,22	8,79	21,7	22,2	-55,6	-18,9
25%	7,72	6,67	9,25	8,88	22	21,7	-52,6	8,2	7,67	7,01	9,23	8,63	21,7	21	-52,2	-19
50%	7,71	6,74	9,19	8,85	22,2	23,2	-52,8	-4,9	7,67	7,09	9,23	8,51	21,9	21,7	-51,3	-14,7
100%	7,61	6,1	9,22	8,78	22,3	25	-48,9	-4,3	7,61	6,75	9,21	8,5	22,2	21,4	-49,6	-13,1
AVERAGE	7,7075	6,6075	9,225	8,81	22,05	23,725	-51,625	-1,325	7,685	6,9725	9,2225	8,6075	21,875	21,575	-52,175	-16,425
DP	0,0741	0,3543	0,0264	0,0678	0,2645	1,5945	1,8337	6,3562	0,0663	0,1519	0,0095	0,1352	0,2362	0,5058	2,5250	2,9881
CV	0,0096	0,0536	0,0028	0,0076	0,0119	0,0672	-0,035	4,7972	0,0086	0,0217	0,0010	0,0157	0,0108	0,0234	-0,0484	-0,18193

Legend: Conc. = Concentration; O.D. = Dissolved Oxygen; Temp. = Temperature; Eh = Redox Potential.

According to the data shown in Table 5.4, it can be concluded that the concentration with the highest toxic potential was 200 mg/L, with equal values for surface and bottom samples (CL_{50} 7.91%). The concentration with the lowest toxic potential was 300 mg/L (CL_{50} 31.5) for surface samples.

The strength of the linear associations between the test results and the physico-chemical parameters was checked using Pearson's linear correlation coefficient. The correlation coefficient is positive if the variables are directly related and negative if they are inversely related, its maximum modular value being equal to 1.

For the toxicological analysis in question, the values marked in red showed significant correlation considering $p > 0.05$. This analysis was used to verify the parameters that contribute most to toxicity, and strong positive correlations were found between percentage dilution and mortality at all sediment concentrations tested (Tables 5.6, 5.7 and 5.8).

Table 5.6 - Pearson's correlation matrix for physico-chemical parameters and toxicity tests for sediment concentration 50 mg/L

50 mg/L	Conc.	Mort.	pH	O.D.	Temp.	Eh
Conc.	1,00					
Mort.	0,93	1,00				
pH	0,97	0,88	1,00			
O.D.	0,39	0,15	0,59	1,00		
Temp.	0,52	0,74	0,56	0,16	1,00	
Eh	-0,90	-0,71	-0,86	-0,41	-0,11	1,00

Legend: Conc. = Concentration; Mort. = Mortality; O.D. = Dissolved Oxygen; Temp. = Temperature; Eh. = Redox Potential.

Table 5.7 - Pearson's correlation matrix for physico-chemical parameters and toxicity tests for sediment concentration 200 mg/L

200 mg/L	Conc.	Mort.	pH	O.D.	Temp.	Eh
Conc.	1,00					
Mort.	0,82	1,00				
pH	-0,96	-0,93	1,00			
O.D.	0,86	0,45	-0,74	1,00		
Temp.	0,96	0,78	-0,90	0,78	1,00	
Eh	0,76	0,95	-0,90	0,46	0,63	1,00

Legend: Conc. = Concentration; Mort. = Mortality; O.D. = Dissolved Oxygen; Temp. = Temperature; Eh = Redox Potential.

Table 5.8 - Pearson's correlation matrix for physico-chemical parameters and toxicity tests for sediment concentration 300 mg/L

300 mg/L	D.P.	Mort.	pH	O.D.	Temp.	Eh
Conc.	1,00					
Mort.	0,98	1,00				
pH	0,97	0,93	1,00			
O.D.	0,39	0,33	0,59	1,00		
Temp.	0,52	0,37	0,56	0,16	1,00	
Eh	-0,90	-0,96	-0,86	-0,41	-0,11	1,00

Legend: Conc. = Concentration; Mort. = Mortality; O.D. = Dissolved Oxygen; Temp. = Temperature; Eh = Redox Potential.

Pearson's correlation analysis (PCA) showed a significant correlation between pH and Eh (oxidation-reduction potential) at all concentrations, with both positive and negative influences depending on the sediment concentration used.

At a concentration of 200 mg/L, these variables behave in an anomalous way, with pH having a strong negative influence and temperature having a positive influence on the increase in toxicity. It is important to note that the redox potential is directly linked to the bioavailability of toxic compounds in the environment, especially metals associated with estuary sediments (NIZOLI; LUIZ-SILVA, 2009).

Principal Component Analysis was used to estimate the similarity of the data, representing it in two orthogonal dimensions. The two factors shown in Figure 5.11, 5.12 and 5.13 explain the largest proportion of the total variance among all the linear combinations of the original data, with the second factor having a smaller proportion of total variance than the first.

For the sediment concentration of 50 mg/L, there was a strong positive correlation between the toxicity factors (percentage dilution and mortality), with dissolved oxygen being the least influential of the physico-chemical parameters monitored. It is also possible to see the strong negative correlation of Eh with the toxicity factors and the other parameters, corroborating the values presented in the Pearson correlation.

Figura 5.11 - Graphical representation of the principal axes (1 and 2) of the principal component analysis of the correlation matrix (percentage dilutions, mortality and physico-chemical parameters) for the sediment concentration 50 mg/L

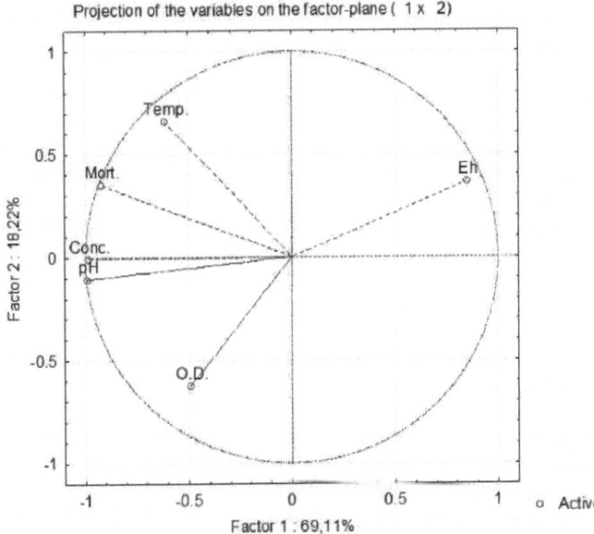

Figura 5.12 - Graphical representation of the principal axes (1 and 2) of the principal component analysis of the correlation matrix (percentage dilutions, mortality and physico-chemical parameters) for the sediment concentration 200 mg/L

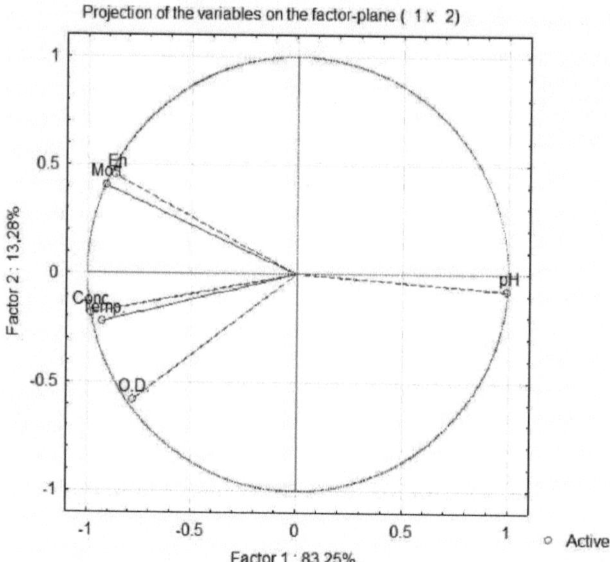

The graph representing the Principal Component Analysis for the 200 mg/L concentration once again shows a strong, positive correlation between the toxicity factors, with pH being the parameter that has an inverse influence on the result, as shown in the Pearson Correlation table. The toxicity factors (concentration and mortality) are once again close to temperature and dissolved oxygen as factors with a strong influence on the results.

Figure 5.13 - Graphical representation of the principal axes (1 and 2) of the principal component analysis of the correlation matrix (percentage dilutions, mortality and physico-chemical parameters) for the sediment concentration 300 mg/L

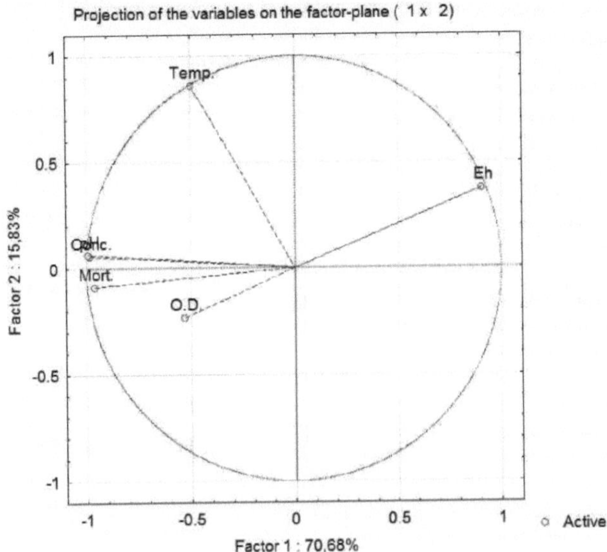

Principal component analysis allowed the data to be represented in a bivariate area. For the toxicological analyses considering the sediment concentration of 50 mg/L, the two main axes - Factor 1 and 2, explain 87.33% of the total variance (18.22% and 69.11%). For the sediment concentration of 200 mg/L, 96.53% of the total variance of the data is explained by the first two factors (83.25% and 13.28%) and at 300 mg/L, the 86.51% variance of the data is explained by factor 1 and 2 at 70.68% and 15.83%, respectively.

In this graphical representation of the test corresponding to a sediment concentration of 300 mg/L, it is possible to corroborate the negative relationship between the oxidation-reduction potential and the positive correlation between the toxicity factors (percentage dilution and mortality) and the other parameters monitored, verified through Pearson's correlation. Once again, dissolved oxygen was considered a factor of low influence for the analysis.

These results corroborate the relationships found between physical-chemical parameters and toxicity factors in previous studies (AMARAL, 2012) where toxicity showed a strong relationship with sensitivity to physical-chemical parameters, observing a negative correlation between pH and biochemical oxygen demand.

With regard to the test's explainability factors, the *eigenvalues* represent the ratio of the variation between the groups to the variation within them, considering that the further away from 1, the greater the variation between the groups explained by the discriminant function (Tables 5.9, 5.10, 5.11).

Table 5.9 - *Eigenvalues*, total percentage of variance and accumulated explanatory factors found for the correlation matrix for sediment concentration 50 mg/L

	Eigenvalues	% total variance	Cumulative eigenvalue	Cumulative %
Component 1	4,146636	69,11060	4,146636	69,1106
Component 2	1,093035	18,21725	5,239671	87,3279
Component 3	0,760329	12,67215	6,000000	100,0000

Table 5.10 - *Eigenvalues*, total percentage of variance and accumulated explanatory factors found for the correlation matrix for sediment concentration 200 mg/L

	Eigenvalues	% total variance	Cumulative eigenvalue	Cumulative %
Component 1	4,995278	83,25463	4,995278	83,2546
Component 2	0,796805	13,28009	5,792083	96,5347
Component 3	0,207917	3,46528	6,000000	100,0000

Table 5.11 - *Eigenvalues*, total percentage of variance and accumulated explanatory factors found for the correlation matrix for sediment concentration 300 mg/L

	Eigenvalues	% total variance	Cumulative eigenvalue	Cumulative %
Component 1	4,146636	69,11060	4,146636	69,1106
Component 2	1,093035	18,21725	5,239671	87,3279
Component 3	0,760329	12,67215	6,000000	100,0000

By analyzing Tables 5.9, 5.10 and 5.11, it is possible to discriminate the three main factors that contribute to the total variance between the parameters, with the first component effectively having the greatest inter-group separating power.

Considering the various processes that combine to lead to the observed result, it is possible to associate the toxicity observed in surface samples with results demonstrated in previous reports. Studies evaluating the persistence of oil and the response of the ecosystem fourteen years after the Exxon Valdez tanker accident concluded that the compounds derived from the oil on the surface persisted in surprising quantities in their most toxic form, being sufficiently bioavailable to induce chronic biological exposures, and that the impacts are strictly associated with the existence of surface sediments (PETERSON et al. 2003).

Therefore, the residual concentrations of toxic chemical compounds on the surface, together with the existence of surface sediment, induce indirect effects on the trophic and interaction chains, transmitting impacts that go far beyond acute mortality.

5.4 CONCLUSIONS

The aim of this study was to evaluate the potential toxic nature of the formation of OSA through three microscale simulation experiments using the microcrustacean Artemia salina as a test organism. To this end, a comparative study was carried out with three different concentrations of sediment based on the percentage dilution of its aqueous phase into surface and bottom fractions.

The concentration with the highest toxic potential was 200 mg/L, with equal values for surface and bottom samples (CL50 7.91%). The concentration with the lowest toxic potential was 300 mg/L (CL50 31.5) for surface samples.

A negative relationship was observed between the oxidation-reduction potential and the hydrogenic potential (only in samples with 200 mg/L of sediment) and the other parameters studied. And there was a positive correlation between the toxicity factors (percentage dilution and mortality) and the other parameters monitored.

Given the complexity and variability of hydrocarbons and water-insoluble compounds, it is necessary to emphasize the importance of relating toxicity analyses to the monitoring of physico-chemical parameters, since the confidence limits of the effects are not very small and in one of the tests they could not be calculated.

REFERENCES:

AJIJOLAIYA, L.O. **The Effects of Mineral Size and Concentration on the Formation of Oil-Mineral Aggregates**. 2004. 80f. Master Thesis, Department of Civil Engineering, Faculty of Engineering, Dalhousie University, Halifax, Canada, 2004.

AMARAL, K. G. C. **Correlation between toxicity factor and physico-chemical parameters for treated domestic effluents**. 2012. 105f Dissertation (Master's Degree in Environmental Science and Technology) - Federal Technological University of Paranâ. Curitiba, 2012.

BARRAGAN, O. L. V. **Geochemical characterization of oils from Latin America.** 2012. 109 f. Dissertation (Master's Degree in Geochemistry: Petroleum and Environment) - Postgraduate Program in Geochemistry, Federal University of Bahia. Salvador, 2012.

BAUMGARTEN, M. G. Z.; ROCHA, J. M. B ; NIENCHESKI, L. F. H. **Manual de análisis em Oceanografia Quimica.** Rio Grande: Editora da FURG, 1996. 132 p.

BRAILE, P. M.; CAVALCANTI, J. E. W. A. **Manual de tratamento de águas residuârias industriais.** Sao Paulo: CETESB, 1993. 764p.

COSTA, C. R.; OLIVI; P.; BOTTA, C. M. R.; ESPINDOLA, E. L. G. Toxicity in aquatic environments: discussion and evaluation methods. **Chem. Nova**. Sao Paulo, v. 31, n. 7, p. 1820-1830, 2008.

FIORUCCI, A. R. The importance of dissolved oxygen in aquatic ecosystems. **Quimica Nova na Escola**, n. 22, p. 10 - 16, nov. 2005.

ITOPF. **Fate of Marine Oil Spills:** Technical Information Paper. London, UK. n 2, 2011. Available at: < http://www.itopf.com/information-services/publications/documents/

tip2fateofmarineoilspills.pdf >. Accessed on: 10 Dec. 2013.

KHELIFA, A.; AJIJOLAIYA, L.O.; MACPHERSON, P.; Lee, K.; HILL, P.S.; GHARBI, S.; BLOUIN, M. **Validation of OMA Formation in Cold Brackish and Sea Waters**. In: PROCEEDINGS OF THE TWENTY-EIGHTH ARCTIC AND MARINE OILSPILL PROGRAM TECHNICAL SEMINAR, ENVIRONMENT. Canada, Ottawa, Ontario, p. 527538, 2005b.

KHELIFA, A.; P.S. HILL; K. LEE. **A stochastic model to predict the formation of OilMineral Aggregates.** In: PROCEEDINGS OF THE TWENTY-SIXTH ARCTIC AND MARINE OILSPILL PROGRAM TECHNICAL SEMINAR. Environment Canada, Ottawa, Ontario, vol. 2, p. 893-909, 2003.

KHELIFA, A.; P.S. HILL; K. LEE. **Modelling the effect of sediment size on OMA formation.** In: PROCEEDINGS OF THE TWENTY-SEVENTH ARCTIC AND MARINE OILSPILL PROGRAM TECHNICAL SEMINAR, Environment Canada, Ottawa, Ontario, vol. 1, p. 383-396, 2004.

KHELIFA, A.; P.S. HILL; P. STOFFYN-EGLI; K. LEE. Effects of salinity and clay composition on Oil-Clay Aggregations. **Marine Environmental Research**, v. 59, p. 235-254, 2005a.

KRUMBEIN, W. C.; SLOSS, L. L. Stratigraphy and sedimentation. **Soil Science**, v. 71, n. 5, p. 329-401, 1951.

LE FLOCH, S.; GUYOMARCH, J.; MERLIN, F.X.; STOFYN-EGLI, P.; DIXON, J.; LEE, K. The influence of salinity on oil-mineral aggregate formation. **Spill Science and Technology Bulletin**, v.8, n. 1, p. 65-71, 2002.

LEACH, W.E. Entomostraca, **Dictionaire des Science Naturelles**, v. 14, pg. 524. 1819.

LEE, K.; P. STOFFYN-EGLI; G.H. TREMBLAY; E.H. OWENS; G.A. SERGY; C.C. GUENETTE; R.C. PRINCE. Oil-Mineral Aggregate formation on oiled beaches: natural attenuation and sediment relocation. **Spill Science and Technology Bulletin**, v. 8, n. 3, p. 285-296, 2003.

LEE, K.; STOFFYN-EGLI P. **Characterization of Oil-Mineral Aggregates.** In: PROCEEDINGS OF THE 2001 INTERNATIONAL OIL SPILL CONFERENCE. American Petroleum Institute, Washington, D.C., p. 991-996, 2001.

LEE, K.; STOFFYN-EGLI, P.; WOOD, P.A.; LUNEL, T. **Formation and structure of OilMineral Fines Aggregates in coastal environments**. In: PROCEEDINGS OF THE TWENTY-FIRST ARCTIC AND MARINE OILSPILL PROGRAM TECHNICAL SEMINAR. Environment Canada, Ottawa, Ontario, pp. 911-921, 1998.

MUSCHENHEIM, D.K.; LEE, K. Removal of oil from the sea surface through particulate interactions: review and prospectus, **Spill Science and Technology Bulletin,** v. 8, n. 1, p. 918, 2002.

NIZOLI, É. C.; LUIZ-SILVA, W. The role of acid volatile sulfide in the control of potential metal bioavailability in contaminated sediments from a tropical estuary, southeast Brazil. **Chem. Nova.** v. 32, n. 2, p. 365-372, 2009.

NUNES, B. S.; CARVALHO F. D.; GUILHERMINO L. M.; STAPPEN G. V. Use of the genus *Artemia* in ecotoxity testing. **Environmental Polution,** v. 144, p. 453-462, 2006.

OMOTOSO, O. E.; MUNOZ, V.A.; MIKULA, R.J. Mechanisms of crude oil-mineral interactions, **Spill Science and Technology Bulletin,** v. 8, n. 1, p. 45-54, 2002.

OWENS, E.H. The interaction of fine particles with stranded oil. **Pure Appl. Chem.,** vol. 71 (1), p. 83-93, 1999.

OWENS, E.H.; LEE, K. Interaction of oil and Mineral fines on shorelines: review and assessment. **Marine Pollution Bulletin**, v. 47, p. 397-405, 2003.

PAYNE, J. R.; CLAYTON J. R.; KIRSTEIN, B.E. Oil/Suspended particulate material interactions and sedimentation. **Spill Science and Technology Bulletin**, v. 8, n. 2, p. 201221, 2003.

PETERSON C. H.; RICE S. D.; SHORT J. W.; ESLER D.; BODKIN J. L.; BALLACHEY B. E. Long-term ecosystem response to the Exxon Valdez oil spill. **Science**, v. 302, p. 20822086. 2003.

PINA, P.; BRAUNSCHWEIG, F.; SARAIVA, S.; SANTOS, M.; MARTINS, F.; NEVES, R. The Importance of Physical Processes in the Control of Eutrophication in Estuaries. **Sapientia.** Algarve, 2003. Available at < http://hdl.handle.net/10400.1/40 >. Accessed on: 04 Apr. 2012.

QUEIROZ, A. F. de S.; CELINO, J. J. Mangroves and estuarine ecosystems in Todos os Santos Bay. In: QUEIROZ, A. F. DE S.; CELINO, J. J. **Avaliação de ambientes na Baia de Todos os Santos: aspectos geoquimicos, geofisicos e biológicos.** Salvador - BA: EDUFBA, 2008. p. 39-58.

SORGELOOS, P.; VAR DER WILEN C.; PERSOONE,G. The use of *Artemia* nauplii for toxicity tests - a critical analysis. **Ecotoxicology and Envitomental Safety,** v. 2, p. 249-255, 1978.

STOFFYN-EGLI, P., LEE, K. Formation and characterization of oil-mineral aggregates. **Spill Science and Technology Bulletin**, v. 8, n. 1, p. 31 - 44, 2002.

SUN, J.; ZHENG, X. A review of oil suspended particulate matter aggregation a natural process of cleansing spilled oil in the aquatic environment. **Journal of environmental monitoring : JEM,** vol. 11, p. 10, 1801-1809. 2009.

VEIGA, L.F.; VITAL, N. Acute toxicity tests with the microcrustacean *Artemia* sp. In. NASCIMENTO, I. A.; SOUSA, E. C. P. M.; NIPPER, M. **Métodos em ecotoxicologia marinha:** aplicações para o Brasil. Sao Paulo, p. 111-119. 2002.

6   CONCLUSIONS

The aim of this study was to evaluate the potential toxicity of the formation of OSA in microscale simulation experiments using the benthic copepod *Nitokra* sp. as a test organism, in comparison with the toxicity found for the elutriate formed from the percentage dilution of oil and sediment. In addition, the potential toxicity of OSA formation was described using three microscale simulation experiments with the microcrustacean *Artemia salina* as the test organism, and a comparative study was carried out with three different sediment concentrations (50, 200, 300 mg/L), based on the percentage dilution of its aqueous phase into surface and bottom fractions.

With regard to the objectives achieved in the first study, it was possible to verify that, with regard to the toxic potential for the cepod *Nitokra sp. in* safeguard of the different intrinsic characteristics of each sample, the concentrated aqueous portion of the simulation of OSA formation on a microscale proved to be considerably less toxic (CL_{50} 70.71%) than the elutriate formed from the percentage dilutions of the sediment with oil (CL_{50} 5.59%). It was also seen that, among the correlations that determine the explanatory power of the influence of the monitored variables, the strong and positive correlations between the toxicity factors (concentration and mortality) stand out, as opposed to dissolved oxygen, which showed a strong and negative correlation.

In the second study, it was found that the sediment concentration with the highest toxic potential for the microcrustacean *Artemia salina* in the microscale simulation experiments was 200 mg/L, with similar values for surface and bottom (CL_{50} 7.91%), while the concentration with the lowest toxic potential was 300 mg/L (CL_{50} 31.5%) for surface samples. There was a negative relationship between the oxidation-reduction potential and the hydrogenic potential (only in samples with 200 mg/L of sediment) and the other parameters studied, and a positive correlation between the toxicity factors (percentage dilution and mortality) and the other parameters monitored.

Given the complexity and variability of hydrocarbons and water-insoluble compounds, it is necessary to emphasize the importance of relating toxicity analyses to the monitoring of physico-chemical parameters, since the confidence limits of the effects are not very small and in one of the tests they could not be calculated.

The aim of this research was to develop a methodology for toxicity analysis that would provide a toxicological basis for the future development of a protocol that would determine a potential methodology for toxicological assessment, providing a basis for the application of OSA in areas impacted by petroleum activities, covering various situations that will have relevant effects, both during its implementation and execution. Also noteworthy is the development of a peculiar and specific methodology that will contribute to current remediation techniques; the determination of the best criteria for characterizing oil-suspended aggregates (OSA) as a natural and efficient dispersant in the process of environmental recovery and ecological biomonitoring.

The technological relevance stems from the fact that no protocol, procedure or methodology has been found that is related to the determination of potential toxicological conditions related to the formation of OSA in mangrove areas impacted by oil in Brazil. Methodological procedures describing the performance of toxicological tests were then suggested with regard to their application, which is not available in any existing methodology, and their recognition will be consolidated as a mechanism to be widely studied due to their efficiency and speed of response to oil spills.

The mangrove areas and conservation units that could be affected by possible accidents involving oil spills in the municipality of Canavieiras, on the southern coast of Bahia, were the target of the studies carried out in this project, which, by consolidating the toxicological definitions of the Oil-Suspended Particulate Matter Aggregate, can help to recognize it as a mechanism that facilitates the remediation of these ecological impacts. The artisanal fishing regions, tourist developments and the Canavieiras RESEX represent important social compartments in this area, which will also be positively influenced by the results of this research.

REFERENCES

ABILIO, F. J. P.; RUFFO, T. L. M.; SOUZA, A. H. F. F.; FLORENTINO, H. S.; OLIVEIRA-JUNIOR, E. T.; MEIRELES, B. N.; SANTANA, A. C. D. Benthic macroinvertebrates as bioindicators of the environmental quality of aquatic bodies in the caatinga. **Oecologia Brasiliense**, v. 11, n. 3, p. 397-409, 2007.

AJIJOLAIYA, L. O., HILL, P. S., KHELIFA, A., ISLAM, R. M., LEE, K., Laboratory investigation of the effects of mineral size and concentration on the formation of oil-mineral aggregations. **Marine Pollution Bulletin**, v. 52, p. 920-927, 2006.

APHA. American Public Health Association. **Standard methods for examination of water and wastewater**. 20th ed. Washington, Port City Press, chap. 1, p. 34-38, 2001.

ARAÙJO, R. J. V.; ARAÙJO-CASTRO, C. M. V.; SOUZA-SANTOS, L. P.. **Sensitivity test of Tisbe biminiensis to Sodium Dodecyl Sulfate**. In: CONGRESSO BRASILEIRO DE ECOTOXICOLOGIA, 9, 2006, Sao Pedro - SP, **Anais...** Sao Paulo: USP, 2006.

ARAÙJO-CASTRO, C. M. V. **Standardization and application of the marine copepod Tisbe biminiensis as a test organism in toxicological evaluations of estuarine sediments**. 104f. 2008. Thesis (Doctorate in Oceanography) - Postgraduate Program in Oceanography. Federal University of Pernambuco. Recife, 2008.

ARAÙJO-C ASTRO, C. M. V.; SOUZA-SANTOS, L. P.; TORREIRO, A. G. A. G.; GARCiA, K. S. Sensitivity of the marine benthic copepod *Tisbe biminiensis* (Copepoda: Harpacticoida) to potassium dichromate and sediment particle size. **Brazilian Journal of Oceanography**, v. 57, p. 33-41, 2009.

BAUMGARTEN, M. G. Z.; ROCHA, J. M. B.; NIENCHESKI, L. F. H. **Manual de análisis em Oceanografia Quimica**, Rio Grande: Editora da FURG, 1996. 132 p.

BRAILE, P. M.; CAVALCANTI, J. E. W. A. **Manual de tratamento de águas residuàrias industriais.** Sao Paulo: CETESB, 1993. 764p.

BRAZIL, National Environment Council. Resolution No. 269 of September 14, 2000.

BRITO, E. M.S.; DURAN, R.; GUYONEAUD, R.; GONI-URRIZA, M.; OTEYZA, T. G.; CRAPEZ, M. A. C.; ALELUIA, I.; WASSERMAN, J. C.A. A case study of *in situ* oil contamination in a mangrove swamp (Rio De Janeiro, Brazil). **Marine Pollution Bulletin**, v. 58, n. 3, p. 418-423, 2009.

CASTRO, P. A. R. **Ecotoxicological monitoring of mangrove sediment contaminated with petroleum hydrocarbons.** 2009. 92 f. Dissertation (Master's Degree in Environmental Engineering) - Postgraduate Program in Environmental Engineering, Federal University of Espirito Santo. Vitória, 2009.

COSTA, B. V. M. **Effect of sieving on the toxicity of oil-contaminated sediment submitted to bioremediation treatments**. 2010. 60p. Monograph (Graduation in Biological Sciences) - Federal University of Pernambuco, Recife, 2010.

COSTA, C. R.; OLIVI; P.; BOTTA, C. M. R.; ESPINDOLA, E. L. G. Toxicity in aquatic environments: discussion and evaluation methods. **Chem. Nova.** v. 31, n. 7, p. 18201830, 2008.

CRUZ, J. F.; TRIGÜIS, J. A.; QUEIROZ, A. F. S. **Evaluation of the effectiveness *versus* effects of surfactants in simulated oil spills at sea.** In: RIO OIL & GAS EXPO AND CONFERENCE, 2012. Rio de Janeiro. **Proceedings...** Rio de Janeiro, 2012

EMBRAPA. National Soil Research Center. **Manual of Chemical Analysis of Soils, Plants and Fertilizers**. SILVA, F. C. da coord. Campinas: Embrapa Informàtica Agropecuària; Rio de Janeiro: Embrapa Solos, 2009. 370p.

FERRAZ, M. A.; ZARONI, L. P.; BERGMANN FILHO, T.U.; GASPARRO, M. R. SOUSA, E. C. P. M. **Egg hatching rate of the copepod *Nitokra* sp. in different salinities.** In: BRAZILIAN CONGRESS OF ECOTOXICOLOGY, 11, 2010. Bombinhas, Santa Catarina. **Proceedings...** Bombinhas, Santa Catarina. 2010.

FIORUCCI, A. R. The importance of dissolved oxygen in aquatic ecosystems. **Quimica Nova na Escola**, n. 22, p. 10-16, 2005.

FOLK, R.L. & WARD, W.C. Brazos river bar: a study of significance of grain size parameters. **Journal of Sedimentary Petrology**, v. 27, p. 3-26, 1957.

FURLEY, T. H.; VITALI, M. M.; SOBREIRA, R. G.; ZARONI, L. P. **Acute sensitivity of the copepod *Nitokra* sp. to zinc sulfate at different salinities**. In: BRAZILIAN CONGRESS OF ECOTOXICOLOGY, 11, 2010. Bombinhas, Santa Catarina. **Proceedings...** Bombinhas, Santa Catarina. 2010.

GARCIA, K. S. **Bioavailability and toxicity of contaminants in sediments in the northeastern portion of Todos os Santos Bay.** 2009. 120 f. Thesis (Doctorate in Geosciences - Geochemistry) - Universidade Federal Fluminense, Niterói, 2009.

GESAMP - Joint Group of Experts on the Scientific Aspects of Marine Pollution - Impact of Oil and Related Chemicals and Wastes on the Marine Environment. **GESAMP Reports and Studies**, n. 50, 1993.

GRASSHOFF, K.; KREMLING, K.; EHRHARDT, M. **Methods of Seawater Analysis** 3ed. Florida: Verlage Chemie. 1999. 417p.

GUYOMARCH, J.; LE FLOCH, S.; MERLIN, F. Effect of Suspended Mineral Load, Water Salinity and Oil Type on the Size of Oil-Mineral Aggregates in the Presence of Chemical Dispersant. **Spill Science and Technology Bulletin**, v. 8, p. 95-100, 2002.

ITOPF. Fate of Marine Oil Spills. **Technical Information Paper.** London, UK. n 2, 2011. Available at: < http://www.itopf.com/information-services/publications/ documents/tip2fateofmarineoilspills.pdf >. Accessed on Dec. 10, 2013.

JENNERJAHN, T. C.; ITTEKKOT, V. Organic matter in sediments in the mangrove areas and adjacent continental margins of Brazil: I. Amino acids and hexoamines. **Oceanologica Acta**, v. 20, n. 2, p. 359 - 369, 1996.

KHELIFA, A.; HILL, P. S.; STOFFYN-EGLI, P.; LEE, K. Effects of salinity and clay type on oil-mineral aggregation. **Marine Environmental** Research, v. 59, p. 235-254, 2005.

KHELIFA, A. B.; FIELDHOUSE, Z.; WANG, C.; YANG, M. L.; FINGAS, M. L.; BROWN, C. E.; GAMBLE, L. A laboratory study on formation of Oil-SPM Aggregates using the NIST Standard Reference Material 1941b. In: ARCTIC AND MARINE OILSPILL PROGRAM TECHNICAL SEMINAR, 30. 2007. **Proceedings...**, Ottawa, Ontario: Environment Canada, v. 1, p. 35-48. 2007.

KUSK, K. O.; WOLLENBERGER, L. Towards an internationally harmonized test method for reproductive and developmental effects of endocrine disrupters in marine copepods. **Ecotoxicology**, v. 16, p. 183-195, 2007.

LEACH, W. E. Entomostraca, **Dictionaire des Science Naturelles**, 14, pg. 524. 1819.

LE FLOCH, S.; GUYOMARCH, J.; MERLIN, F.X.; STOFYN-EGLI, P.; DIXON, J.; LEE, K. The influence of salinity on Oil-Mineral Aggregate formation. **Spill Science and Technology Bulletin**, v. 8, n. 1, p.65-71, 2002.

LEE, K.; STOFFYN-EGLI, P. Characterization of oil-mineral aggregates. In: INTERNATIONAL OIL SPILL CONFERENCE. 2001. **Proceedings.** Washington, DC: American Petroleum Institute, p. 991-996, Publication Number 14710, 2001.

LEE, K.; STOFFYN-EGLI, P.; OWENS, E. H. The OSSA II pipeline oil spill: Natural mitigation of a riverine oil spill by oil-mineral aggregate formation. **Spill Science and Technology Bulletin**, v. 7, p.149-154, 2002.

LOTUFO, G. R.; ABESSA, D.M. S. Toxicity tests with total sediment and estuarine interstitial water using benthic copepods. In: NASCIMENTO, I. A.; SOUSA, E. C. P. M.; NIPPER, M. **Métodos em ecotoxicologia marinha: aplicações para o Brasil.** Sao Paulo, 2002. p. 151-162.

MANÇÙ, R. J. S.; MONTEIRO, A. O.; BRUNI, E: L. Environmental management in oil production in the state of Bahia: a comparison between the adherence of the environmental management practices of a national and a foreign company to international standards. In: COLLOQUE DE L'IFBAE, 5. Grenoble, 2009.

MORAES, R. B. C.; CRAPEZ, M. A. C.; PFEIFFER, W. C.; FARINA, M.; BAINY, A. C. D. (Org.); TEIXEIRA, V. L. (Org.). **Efeito de poluentes em organismos marinhos.** 1. ed. Sao Paulo: Arte Ciência Villipress, 2001, v. 1. 285 p.

MOREIRA, I. T. A. **Investigation of possible ecological impacts of oil on estuarine biological communities in Todos os Santos Bay and Southern Bahia: OSA as a guiding tool.** 2014. 265f. Thesis (PhD in Geology) - Institute of Geosciences, Federal University of Bahia, Salvador, 2014.

NASCIMENTO, I. A.; SOUSA, E. C. P. M.; NIPPER, M. **Métodos em ecotoxicologia marinha:** aplicações para o Brasil. Sao Paulo: Artes Gráficas e Ind. Ltda, 2002. 262 p.

NIU, H.; LI, Z.; LEE, K.; KEPKAY, P.; MULLIN, J. V. Modelling the transport of OilMineral-Aggregates (OMAs) in the marine environment and assessment of their potential risks. **Environmental Modeling and Assessment**, v. 16, p. 61-75, 2011.

NUNES, B. S.; CARVALHO F. D.; GUILHERMINO L. M.; STAPPEN G. V. Use of the genus *Artemia* in ecotoxity testing. **Environmental Pollution**, v. 144, p. 453-462. 2006.

OLIVEIRA, D. D. **Evaluation of the sensitivity of the copepod** *Tisbe biminiensis to* **toxicological tests in estuarine water and sediments.** 2011. 82 f Dissertation (Master's Degree in Oceanography) - Federal University of Pernambuco, Recife, 2011.

OWENS, E. H., LEE, K. Interaction of oil and mineral fines on shorelines: review and assessment. **Marine Pollution Bulletin**, v.47, p. 397-405. 2003.

PEDROZO, M. F. M.; BARBOSA, E. M.; COURSEIL, H. X.; SCHNEIDER, M. R.; LINHARES, M. M. **Ecotoxicologia e Avaliação de risco do Petróleo.** Série de Cadernos de Referência Ambiental: CRA, Salvador. v. 12, 2002, 246 p.

PETROBRAS E. and P. **Modeling the transport and dispersion of oil at sea for Block BMJ-1 Jequitinhonha Basin**, Nov. 2007.

PIMENTEL, F. P. **Analysis of response strategies to oil spills in the Golfinho field (ES - Brazil) using the Oscar model.** 2007. 109 f. Monograph (Graduation in Oceanography) - Federal University of Espirito Santo, Vitória, 2007.

PINA, P.; BRAUNSCHWEIG, F.; SARAIVA, S.; SANTOS, M.; MARTINS, F.; NEVES, R. The Importance of Physical Processes in the Control of Eutrophication in Estuaries. **Sapientia.** Algarve, 2003. Available at < http://hdl.handle.net/10400.1/40 >. Accessed on: 04 Apr. 2012.

QUEIROZ GALVÂO PERFURAÇÔES S. A. **Maritime Drilling Activity in Block BM-J-2: Environmental Impact Study - EIA**, Item II-9, Feb. 2006.

QUEIROZ, A. F. de S.; CELINO, J. J. Mangroves and estuarine ecosystems in Todos os Santos Bay. In: QUEIROZ, A. F. DE S.; CELINO, J. J. **Avaliação de ambientes na Baia de Todos os Santos: aspectos geoquimicos, geofisicos e biológicos.** Salvador - BA: EDUFBA, 2008. p. 39-58.

QUEIROZ, A. F. S.; CELINO, J. J. Environmental impact of the oil industry on mangroves in the northern region of Todos os Santos Bay (Bahia, Brazil). **Boletim Paranaense de Geociências**, n. 62-63, p. 23-34, 2008.

RODRIGUES, A. J. C. **Characterization of oil-mineral particle aggregates: procedures for accelerating the remediation of oil spills in coastal environments.** 2011. 54f. Monograph (Graduation in Geology) - Federal University of Bahia, Salvador, 2011.

SAMPAIO, S. C.; SILVESTRO, M. G.; FRIGO, E. P.; BORGES, C. M. Relationship between solids series and electrical conductivity in different wastewaters. **Irriga**, v. 12, n. 4, p. 557562, 2007.

SANTOS, D. R. **Investigation of oil-mineral aggregate (OMA) interaction in coastal environments under the influence of different salinities: a subsidy for oil spill remediation**

procedures. 2010. 59f. Monograph (Graduation in Geology) - Federal University of Bahia, Salvador, 2010.

SANTOS, J. M. N. **Utilization of benthic invertebrates in sediment ecotoxicology.** 2008. 58f. Dissertation (Master's Degree in Toxicology and Ecotoxicology) - University of Aveiro, Portugal, 2009.

SANTOS, L. C.; CUNHA-LIGNON, M.; SCHAEFFER-NOVELLI, Y.; CITRÓN-MORELO, G. Long-term effects of oil pollution in mangrove forests (Baixada Santista, Southeast Brazil) detected using a GIS-Based multitemporal analysis of aerial photographs. **Brazilian Journal of Oceanography**, v. 60, p. 161-172, 2012.

SANTOS, P. S.; MARQUES, A. C.; ARAÙJO, M. **Remnants of coastal vegetation in the southeast region of Bahia - municipalities of Una and Canavieiras.** In: GIS BRASIL: 2ᵃ MOSTRA DO TALENT CIENTiFICO, 2, Curitiba, 2002. **Proceedings...** Curitiba, 2012.

SCHMIDT, A. J.; OLIVEIRA, M. A. and collaborators - **Action plan for the uçà crab in Canavieiras.** 96p. ALMA Project - Atlantic Forest Coastal Environments, Ecotuba Project, Comandatuba Island, BA, Brazil, 2006.

SERGY, G. A.; GUÉNETTE, C. C.; OWENS, E. H.; PRINCE, R. C.; LEE, K. In-situ Treatment of Oiled Sediment Shorelines. **Spill Science and Technology Bulletin**, v. 8, n. 3, p. 237-244, 2003.

SORGELOOS, P.; VAR DER WILEN C.; PERSOONE, G. The use of *Artemia* nauplii for toxicity tests - a critical analysis. **Ecotoxicology and Envitomental Safety**, v. 2, p. 249-255, 1978.

SOUSA, E. C. P. M.; ZARONI, L. P.; FILHO T. U. B.; MARCONATO, A. A.; KIRSCHBAUM, A. A.; GASPARRO, M. R. Acute sensitivity of *Nitokra* sp benthic copepod to potassium dichromate and ammonia chloride. **J. Braz. Soc. Ecotoxicol.**, v. 7, n. 1, p. 75-81 2012.

STOFFYN-EGLI, P., LEE, K. Formation and characterization of oil- mineral aggregates. **Spill Science and Technology Bulletin**, v. 8, n. 1, p. 31 - 44, 2002.

SUN, J.; ZHENG, X. A review of oil suspended particulate matter aggregation a natural process of cleansing spilled oil in the aquatic environment. **Journal of environmental monitoring: JEM**, v. 11, n. 10, p. 1801-1809. 2009.

TRINDADE, M. C. L. F.; OLIVEIRA, O. M. C.; MOREIRA I. T. A.; QUEIROZ, A. F. S.; SILVA, C. S.; RIOS M. C. **Formation and characterization of oil-mineral aggregate at different hydrodynamic energies: important role in the natural removal of oil spills.** In: RIO OIL & GAS EXPO AND CONFERENCE, 2012. Rio de Janeiro. **Proceedings...** Rio de Janeiro, 2012

UNITED STATES ENVIRONMENTAL PROTECTION AGENCY - USEPA. Testing Manual

EPA-821-R-02-012 - **Methods for Measuring the Acute Toxicity of Effluents and Receiving Waters to Freshwater and Marine Organisms.** Oct. 2002. Available at: < http://water.epa.gov/scitech/methods/cwa/wet/upload/2007_07_10_methods_wet_disk2_atx1- 6.pdf '. Accessed on 10 Jan 2014.

UNITED STATES ENVIRONMENTAL PROTECTION AGENCY - USEPA. **Testing Manual 503/8-91/001 - Evaluation of Dredged Material Proposed for Ocean Disposal.** 1991. Feb1991 . Available :<

http://water.epa.gov/type/oceb/oceandumping/dredgedmaterial/upload/gbook.pdf '. Accessed 10 Dec 2013.

UNITED STATES ENVIRONMENTAL PROTECTION AGENCY - USEPA. **Testing Manual EPA-823-B-01-002 - Methods for Collection, Storage, and Manipulation of Sediments for Chemical and Toxicological Analyses.** Oct 2001. Available at: < http://water.epa.gov/polwaste/sediments/cs/upload/collectionmanual.pdf '. Accessed on 04 Feb 2014.

VEIGA, L.F.; VITAL, N. Acute toxicity tests with the microcrustacean *Artemia* sp. In. NASCIMENTO, I. A.; SOUSA, E. C. P. M.; NIPPER, M. **Métodos em ecotoxicologia marinha:** aplicações para o Brasil. Sao Paulo: Artes Gráficas e Ind. Ltda, 2002. 262 p.

VENTURINI, E.; TOMMASI, L. R.; BiCEGO, M. C.; MARTINS, C. C. Characterization of the benthic environment of a coastal area adjacent to an oil refinery, Todos os Santos bay (ne- Brazil). **Brazilian Journal of Oceanography**, v. 52, n. 2, p. 123 - 134, 2004.

VOLKMANN-ROCCO, B. Tisbe biminiensis (Copepoda, Harpacticoida) new species of the gracilis group. **Archives Oceanography Limnology**, v. 8, p. 71-90, 1973.

WALKEY-BLACK, A. A critical examination of a rapid method for determining organic carbon in soils: Effect of variations in digestion conditions and of inorganic soil constituents. **Soil Science**, v. 63, p. 251- 263, 1947.

WANG, W.; ZHENG, Y; LEE, K. Chemical dispersion of oil with mineral fines in a low temperature environment. **Marine Pollution Bulletin**, v. 72, p. 205 - 2012, 2013.

WILLIAMS, T. D. Survival and development of copepod larvae Tisbe battagliai in surface microlayer water and sediment elutriates from the German Bight. **Marine Ecology Progress Series**, v. 91, p. 221-228, 1992.

WILLIAMS, T. D.; JONES, M. B. Effects of temperature and food quantity on postembryonic development of *Tisbe battagliai* (Copepoda: Harpacticoida). **Journal Experimental Marine**

Biology and Ecology, v. 183, p. 283-298, 1994.

ZARONI, L. P.; BERGMANN FILHO, T.U.; FERRAZ, M. A.; GASPARRO, M. R. &SOUSA, E. C. P. M. **Egg hatching rate of the copepod *Nitokra* sp. in different types of sediment.** In: BRAZILIAN CONGRESS OF ECOTOXICOLOGY, 11, 2010. Bombinhas, Santa Catarina. **Proceedings...** Bombinhas, Santa Catarina. 2010.

ZIOLLI, R. L. Environmental aspects involved in marine oil pollution. **Magazine**

Saùde e Ambiente / Health and Environment Journal, v.3, n. 2, p. 32-36, 2002.

7 APPENDICES

APPENDIX A - Theoretical framework

Aggregates Oil-Suspended Particulate Matter (OSA)

Oil-Suspended Particulate Aggregates (OSA) are microscopic entities composed of oil and agglutinated minerals that are stable for long periods in brackish and saline waters (LE FLOCH et al., 2002). They are formed very easily in the environment from the interaction of fine mineral particles and oil, and are considered to be effective mineral dispersants (STOFFYN-EGLI; LEE, 2002).

The most common type of OSA is the *droplet-type* aggregate, where a single drop of oil establishes an aggregation with mineral particles attached to its surface. Sometimes, more than one droplet can be present in a total of multiple droplets. In the *solid* type of aggregates, minerals are also present within the oil, forming non-spherical shapes. Another type is *flakes,* which have the appearance of membranous structures and are generally floating, reaching several millimeters in length. OSA can be positively or negatively buoyant depending on its oil to mineral volume ratio (LEE; STOFFYN-EGLI, 2001; STOFFYN-EGLI; LEE, 2002).

Studies suggest that these aggregates are affected by various factors, such as oil and sediment characteristics (mineralogy, granulometry, organic matter content), turbulence, salinity and temperature (AJIJOLAIYA et al., 2006; KHELIFA et al., 2002; 2005; STOFFYN-EGLI; LEE, 2002). The formation of OSA has been considered a natural cleaning process for oiled coastlines, as it increases the rate of dispersion and biodegradation of the oil, as well as preventing it from adhering to the sediment (LEE, et al., 1997; OWENS, 1999).

Among the various environmental factors that determine the formation of this aggregate, some of the main ones are worth highlighting. The effect of temperature on the formation of OSA is essentially the result of changes in the oil's viscosity, establishing an inverse relationship. Lower temperatures reduce the formation of OSA, since viscosity increases with decreasing temperature (OWENS; LEE, 2003; KHELIFA et al., 2002). The type of suspended sediment, its concentration, granulometry and mineralogical composition are also factors that control the rate of formation and shape of OSA (AJIJOLAIYA et al., 2006; GUYOMARCH et al., 2002; STOFFYN-EGLI; LEE, 2002).

The energy of the environment is also a fundamental factor limiting the formation of OSA. The level and duration of the turbulent energy of the environment, associated with the action of waves and currents, is a key factor in the formation of aggregates, increasing their rate of formation. The breakdown of the surface tension of the oil slick accelerates the formation of droplets, the high energy acts as an inducer for the resuspension of the sediment and its maintenance in the water column, promoting collisions between the oil droplets and the sediment (SERGY et al., 2003; NIU et al., 2010).

Salinity has a strong effect on the characteristics of mineral-stabilized oil droplets and is a prerequisite for flocculation between oil and suspended particulate matter. The concentration of oil droplets increases rapidly as salinity rises. There is a critical aggregation salinity at which the rapid increase in OSA formation stops and the stabilized droplets reach their maximum value (OWENS; LEE, 2003).

In Brazil, studies have been carried out to characterize and adapt the use of this technique to specific local conditions. Among the main studies, we can highlight the research carried out by Santos (2010), which aimed to understand the effect of salinity on the formation of OSA in estuarine areas of the Bay of All Saints, through laboratory tests. These studies concluded that oil degradation is favored by the formation of OSA, and is applicable to both marine and estuarine waters.

Rodrigues (2011) has also carried out studies, also at laboratory level, which have made it possible to subsidize the use of OSA, facilitating an understanding of the formation of this aggregate from exposure to different granulometric fractions of the sediment, and how these interfere in its formation. The author concluded that in tropical marine and estuarine sediment environments with sand, silt and clay granulometric characteristics, the formation of OSA could favor the degradation of spilled oils in both marine and estuarine waters and strongly depends on the presence of associated clay minerals.

Correlating the different hydrodynamic energies and mineralogical assemblages, Trindade and colleagues (2013) concluded that energy is a determining factor in the formation of OSA, and is also influenced by the mineralogical composition that defines the variety of size and shape (type) of the aggregates, based on laboratory simulations.

Toxicology

Toxicity tests can be defined as laboratory tests which, carried out under specific and controlled experimental conditions, help to estimate the toxicity of different substances, industrial effluents and environmental samples (water or sediment). Toxicological tests are based on exposing test organisms to different concentrations of the sample to be analyzed, and the toxic effects exhibited are observed, described and quantified. Describing the toxic effects of the substances analyzed on marine and estuarine organisms is essential for maintaining biodiversity (NASCIMENTO; SOUZA; NIPPER, 2002).

The first records of toxicity tests with aquatic organisms date back to 1920, with fish being used first. In the 1940s and 1950s, work in the field increased and other methods emerged, as well as differences in experimental conditions. The first studies related to environmental toxicology included pesticide action tests on insects and water quality tests with cladocerans. In the 1940s and 1950s, studies began on the effects of chemical substances and residues on the environment, and with them the realization that these tests needed to be standardized (HUNN, 1989 apud NASCIMENTO; SOUZA; NIPPER, 2002).

In general, toxicity tests must be regulated by bodies that develop test protocols and technical standards. In Brazil, the body responsible for developing these protocols is the Brazilian Association of Technical Standards (ABNT). The São Paulo State Environmental Sanitation Technology Company (CETESB) also standardizes toxicity tests (Table 1). Other international bodies such as *Association Française de Normalisation* (AFNOR), *American Society for Testing and Materials* (ASTM), *American Water Work Association* (AWWA), *Deutsches Institut fur Normung* (DIN), *International Organization for Standardization* (ISO) and *Organization for Economic Cooperation and Development* (OECD) have also focused on the development of toxicity testing protocols that define permissible toxicity limits with acceptable levels of uncertainty and serve as a reference for regulatory bodies when making decisions (ARAGÂO and ARAÙJO, 2006; COSTA et al., 2008).

Table 1 - Toxicity tests standardized by ABNT and CETESB

Organization	Effect	Species	Brazilian standards
Bacteria	Acute	*Vibrio fischeri*	CETESB, L5.22740
Bacteria	Acute	*Spirillum volatans*	CETESB, L5.22841
Algae	Chronic	*Chlorella vulgaris; Scenedesmus subspicatus; Pseudokirchneriella subcaptata.*	CETESB, L5.02042 and ABNT NBR 1264843
Microcrustacean	Acute	*Daphnia smilis;Daphnia magna*	CETESB, L5.01844 and ABNT NBR 1271345
Microcrustacean	Acute	*Artemia salina*	CETESB, L5.02146
Microcrustacean	Chronic	*Ceriodaphinia dùbia; Ceriodaphninia silvestrii*	CETESB, L5.02247 and ABNT NBR 1337348
Fish	Acute	*Danio rerio; Pimephales promelias*	CETESB, L5.01949 and

			ABNT NBR 1508850

Source: adapted from COSTA et al., (2008).

Toxicity tests are classified into acute and chronic. Chronic toxicity tests are used to estimate the toxic effects of substances under conditions of prolonged exposure to sub-lethal concentrations, i.e. concentrations that do not limit the survival of organisms, but which in some way affect their biological functions in conditions such as mobility, reproduction, feeding or other metabolic functions. The results obtained in chronic toxicity tests are generally expressed as CENO (Concentration of Unobserved Effect) or CEO (Concentration of Observed Effect), but can also be expressed as CE_{50} (Average Effective Concentration in 50% of organisms) (ARAÙJO; ARAÙJO-CASTRO; SOUZA-SANTOS, 2006; KUSK; WOLLENBERGER, 2007; ARAUJO-CASTRO, 2008; ARAUJO-CASTRO et al, 2009; GARCIA, 2009; OLIVEIRA, 2011).

Acute toxicity tests are widely used to measure the possible effects of toxic substances on certain species over a short period of time compared to the life span of the test organism in question. It assesses the dose or concentration of a toxic agent that would produce a specific measurable reaction in a test organism over a short period of time, usually 24 to 96 hours. The main effect measured in acute toxicity studies is lethality or some other behavior that precedes it, such as immobility. Acute toxicity tests allow for EC50 (Effective Concentration) and LC50 (Lethal Concentration) values. The values of effective and lethal concentrations are always related to 50% of the organisms because these responses are more reproducible, can be estimated with a greater degree of reliability and are more significant for extrapolation to a population (COSTA et al., 2008).

APPENDIX B - Study Area

The municipality of Canavieiras (Figure 1) is located in the southern region of the state of Bahia, at 15° 41' south latitude and 38° 57' west longitude (SANTOS et al., 2002). It has an area of 1390 km^2 where there are 7,404ha of economically exploited mangroves within the Canavieiras Extractive Reserve (RESEX), which encompasses most of the estuarine region (101,000ha), supporting around 37,000 inhabitants (SCHMIDT et al., 2006).

Figura 1   - Location map of the study area

Source: adapted from QUEIROZ GALVÂO PERFURAÇÔES S. A., (2006).

Block BM-J-2, with 317 km^2 , which was awarded by the National Petroleum Agency (ANP) to Queiroz Galvao Exploraçao e Produçao S.A., is located in the Jequitinhonha Sedimentary Basin, on the southern coast of the state of Bahia, between the tip of Itapua (Ilhéus) and the island of Atalaia (Canavieiras), about 20 km from the coast. In this location, the oil exploration phase has been carried out at depths of around 4,700 m in a water depth of approximately 44 meters (Petrobràs, 2007).

Taking into account the risks associated with this operation, in 2006 Queiroz Galvao Perfuraçoes S. A. prepared an Individual Emergency Plan (PEI) for exploratory drilling activities. In this study, the municipality of Canavieiras is indicated in all the probabilistic simulations as the most likely to be affected in the event of accidents in which the oil could reach the coast, affecting large areas of mangroves and conservation units, artisanal fishing regions and tourist developments, justifying the choice of this municipality for the scope of this study (Figure 2).

Figura 2 - Probabilistic model of contact with the stain of a possible leak

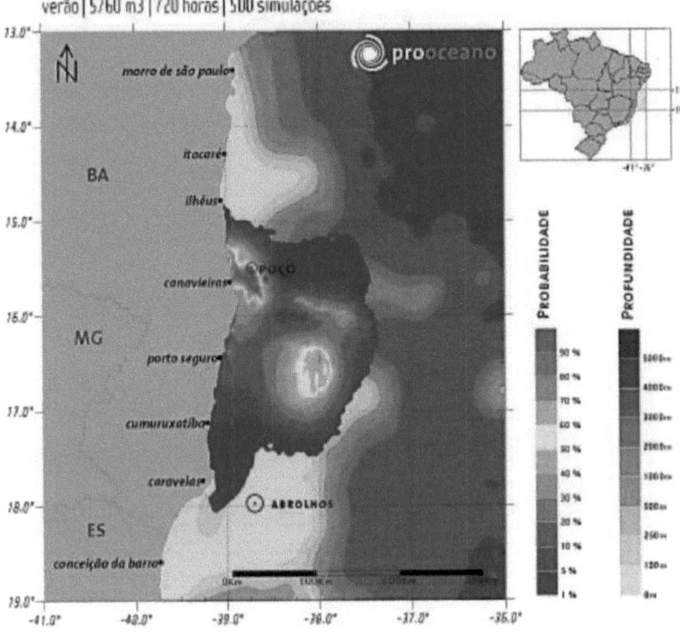

Source: QUEIROZ GALVÂO PERFURAÇOES S. A., (2006).

APPENDIX C - Particle size characterization of the sediment along the Rio Pardo in the municipality of Canavieiras, southern coastal region of the state of Bahia, at the different sampling points

Sample/sediment	% Coarse sand	% Medium sand	% Fine sand	% Very fine sand	% Silt	% Clay
P1.1 (MANG/CAV)	14,58%	0,00%	3,08%	44,25%	37,41%	0,69%
P1.2 (MANG/CAV)	9,93%	0,00%	6,30%	46,98%	36,30%	0,52%
P1.3 (MANG/CAV)	6,40%	0,00%	2,06%	46,70%	44,18%	0,66%
P2.1 (MANG/CAV)	-	-	-	-	-	-
P2.2 (MANG/CAV)	19,68%	0,00%	0,26%	35,50%	43,53%	1,00%
P2.3 (MANG/CAV)	18,76%	0,00%	0,17%	32,54%	47,44%	1,11%
P3.1 (MANG/CAV)	18,64%	0,00%	4,21%	46,11%	30,59%	0,46%
P3.2 (MANG/CAV)	17,10%	0,00%	2,48%	46,24%	33,66%	0,56%
P3.3 (MANG/CAV)	18,89%	0,00%	2,79%	44,82%	32,93%	0,56%
P4.1 (MANG/CAV)	30,30%	0,00%	2,88%	37,14%	29,00%	0,67%
P4.2 (MANG/CAV)	21,28%	0,00%	4,07%	41,66%	32,27%	0,69%
P4.3 (MANG/CAV)	18,80%	0,00%	4,60%	43,36%	32,64%	0,63%
P5.1 (MANG/CAV)	26,35%	0,00%	2,61%	35,94%	34,17%	0,92%
P5.2 (MANG/CAV)	24,25%	0,00%	0,88%	34,50%	39,24%	1,13%

P5.3 (MANG/CAV)	25,02%	0,00%	2,02%	35,31%	36,74%	0,90%
P6.1 (MANG/CAV)	25,32%	0,00%	0,26%	34,02%	39,35%	1,08%
P6.2 (MANG/CAV)	31,30%	0,00%	0,29%	31,22%	36,20%	1,01%
P6.3 (MANG/CAV)	29,96%	0,00%	0,41%	33,73%	35,02%	0,86%

Buy your books fast and straightforward online - at one of world's fastest growing online book stores! Environmentally sound due to Print-on-Demand technologies.

Buy your books online at
www.morebooks.shop

Kaufen Sie Ihre Bücher schnell und unkompliziert online – auf einer der am schnellsten wachsenden Buchhandelsplattformen weltweit! Dank Print-On-Demand umwelt- und ressourcenschonend produzi ert.

Bücher schneller online kaufen
www.morebooks.shop

Printed by Books on Demand GmbH, Norderstedt / Germany